ADVANCES in the
NEUROSCIENCE of ADDICTION

FRONTIERS IN NEUROSCIENCE

Series Editors
Sidney A. Simon, Ph.D.
Miguel A.L. Nicolelis, M.D., Ph.D.

Published Titles

The Superior Colliculus: New Approaches for Studying Sensorimotor Integration
William C. Hall, Ph.D., Department of Neuroscience, Duke University,
    Durham, North Carolina
Adonis Moschovakis, Ph.D., Department of Basic Sciences, University of Crete,
    Heraklion, Greece

New Concepts in Cerebral Ischemia
Rick C. S. Lin, Ph.D., Professor of Anatomy, University of Mississippi Medical Center,
    Jackson, Mississippi

DNA Arrays: Technologies and Experimental Strategies
Elena Grigorenko, Ph.D., Technology Development Group, Millennium Pharmaceuticals,
    Cambridge, Massachusetts

Methods for Alcohol-Related Neuroscience Research
Yuan Liu, Ph.D., National Institute of Neurological Disorders and Stroke,
    National Institutes of Health, Bethesda, Maryland
David M. Lovinger, Ph.D., Laboratory of Integrative Neuroscience, NIAAA,
    Nashville, Tennessee

Primate Audition: Behavior and Neurobiology
Asif A. Ghazanfar, Ph.D., Princeton University, Princeton, New Jersey

Methods in Drug Abuse Research: Cellular and Circuit Level Analyses
Barry D. Waterhouse, Ph.D., MCP-Hahnemann University, Philadelphia, Pennsylvania

Functional and Neural Mechanisms of Interval Timing
Warren H. Meck, Ph.D., Professor of Psychology, Duke University, Durham, North Carolina

Biomedical Imaging in Experimental Neuroscience
Nick Van Bruggen, Ph.D., Department of Neuroscience Genentech, Inc.
Timothy P.L. Roberts, Ph.D., Associate Professor, University of Toronto, Canada

The Primate Visual System
John H. Kaas, Department of Psychology, Vanderbilt University
Christine Collins, Department of Psychology, Vanderbilt University, Nashville, Tennessee

Neurosteroid Effects in the Central Nervous System
Sheryl S. Smith, Ph.D., Department of Physiology, SUNY Health Science Center,
    Brooklyn, New York

Modern Neurosurgery: Clinical Translation of Neuroscience Advances
Dennis A. Turner, Department of Surgery, Division of Neurosurgery,
    Duke University Medical Center, Durham, North Carolina

Sleep: Circuits and Functions
Pierre-Hervé Luppi, Université Claude Bernard Lyon, France

Methods in Insect Sensory Neuroscience
Thomas A. Christensen, Arizona Research Laboratories, Division of Neurobiology,
    University of Arizona, Tuscon, Arizona

Motor Cortex in Voluntary Movements
Alexa Riehle, INCM-CNRS, Marseille, France
Eilon Vaadia, The Hebrew University, Jerusalem, Israel

Neural Plasticity in Adult Somatic Sensory-Motor Systems
Ford F. Ebner, Vanderbilt University, Nashville, Tennessee

Advances in Vagal Afferent Neurobiology
Bradley J. Undem, Johns Hopkins Asthma Center, Baltimore, Maryland
Daniel Weinreich, University of Maryland, Baltimore, Maryland

The Dynamic Synapse: Molecular Methods in Ionotropic Receptor Biology
Josef T. Kittler, University College, London, England
Stephen J. Moss, University College, London, England

Animal Models of Cognitive Impairment
Edward D. Levin, Duke University Medical Center, Durham, North Carolina
Jerry J. Buccafusco, Medical College of Georgia, Augusta, Georgia

The Role of the Nucleus of the Solitary Tract in Gustatory Processing
Robert M. Bradley, University of Michigan, Ann Arbor, Michigan

Brain Aging: Models, Methods, and Mechanisms
David R. Riddle, Wake Forest University, Winston-Salem, North Carolina

Neural Plasticity and Memory: From Genes to Brain Imaging
Frederico Bermudez-Rattoni, National University of Mexico, Mexico City, Mexico

Serotonin Receptors in Neurobiology
Amitabha Chattopadhyay, Center for Cellular and Molecular Biology, Hyderabad, India

TRP Ion Channel Function in Sensory Transduction and Cellular Signaling Cascades
Wolfgang B. Liedtke, M.D., Ph.D., Duke University Medical Center, Durham, North Carolina
Stefan Heller, Ph.D., Stanford University School of Medicine, Stanford, California

Methods for Neural Ensemble Recordings, Second Edition
Miguel A.L. Nicolelis, M.D., Ph.D., Professor of Neurobiology and Biomedical Engineering,
    Duke University Medical Center, Durham, North Carolina

Biology of the NMDA Receptor
Antonius M. VanDongen, Duke University Medical Center, Durham, North Carolina

Methods of Behavioral Analysis in Neuroscience
Jerry J. Buccafusco, Ph.D., Alzheimer's Research Center, Professor of Pharmacology
    and Toxicology, Professor of Psychiatry and Health Behavior,
    Medical College of Georgia, Augusta, Georgia

***In Vivo* Optical Imaging of Brain Function, Second Edition**
Ron Frostig, Ph.D., Professor, Department of Neurobiology,
    University of California, Irvine, California

Fat Detection: Taste, Texture, and Post Ingestive Effects
Jean-Pierre Montmayeur, Ph.D., Centre National de la Recherche Scientifique, Dijon, France
Johannes le Coutre, Ph.D., Nestlé Research Center, Lausanne, Switzerland

The Neurobiology of Olfaction
Anna Menini, Ph.D., Neurobiology Sector International School for Advanced
    Studies,(S.I.S.S.A.), Trieste, Italy

Neuroproteomics
Oscar Alzate, Ph.D., Department of Cell and Developmental Biology,
    University of North Carolina, Chapel Hill, North Carolina

Translational Pain Research: From Mouse to Man
Lawrence Kruger, Ph.D., Department of Neurobiology, UCLA School of Medicine,
    Los Angeles, California
Alan R. Light, Ph.D., Department of Anesthesiology, University of Utah,
    Salt Lake City, Utah

Advances in the Neuroscience of Addiction
Cynthia M. Kuhn, Duke University Medical Center, Durham, North Carolina, USA
George F. Koob, The Scripps Research Institute, La Jolla, California, USA

ADVANCES in the NEUROSCIENCE of ADDICTION

Edited by
Cynthia M. Kuhn
Duke University Medical Center
North Carolina

George F. Koob
Scripps Research Institute
California

CRC Press
Taylor & Francis Group
Boca Raton   London   New York

CRC Press is an imprint of the
Taylor & Francis Group, an **informa** business

CRC Press
Taylor & Francis Group
6000 Broken Sound Parkway NW, Suite 300
Boca Raton, FL 33487-2742

First issued in paperback 2017

ISBN 13: 978-1-138-11642-9 (pbk)
ISBN 13: 978-0-8493-7391-6 (hbk)

Library of Congress Cataloging-in-Publication Data

Advances in the neuroscience of addiction / editors, Cynthia M. Kuhn, George F. Koob.
    p. ; cm. -- (Frontiers in neuroscience)
    Includes bibliographical references and index.
    ISBN 978-0-8493-7391-6 (hardcover : alk. paper)
    1. Substance abuse--Physiological aspects. 2. Neurosciences. I. Kuhn, Cynthia. II. Koob, George F. III. Series: Frontiers in neuroscience (Boca Raton, Fla.)
    [DNLM: 1. Substance-Related Disorders--metabolism. 2. Behavior, Addictive--metabolism. 3. Research. WM 270 A2435 2010]

RC564.A327 2010
616.86--dc22
                                                               2009039060

Visit the Taylor & Francis Web site at
http://www.taylorandfrancis.com

and the CRC Press Web site at
http://www.crcpress.com

Contents

Series Preface

Our goal in creating the Frontiers in Neuroscience Series is to present the insights of experts on emerging fields and theoretical concepts that are, or will be, in the vanguard of neuroscience. Books in the series cover genetics, ion channels, apoptosis, electrodes, neural ensemble recordings in behaving animals, and even robotics. The series also covers new and exciting multidisciplinary areas of brain research, such as computational neuroscience and neuroengineering, and describes breakthroughs in classical fields like behavioral neuroscience. We hope every neuroscientist will use these books in order to get acquainted with new ideas and frontiers in brain research. These books can be given to graduate students and postdoctoral fellows when they are looking for guidance to start a new line of research.

Each book is edited by an expert and consists of chapters written by the leaders in a particular field. Books are richly illustrated and contain comprehensive bibliographies. Chapters provide substantial background material relevant to the particular subject. We hope that as the volumes become available, the effort put in by us, the publisher, the book editors, and individual authors will contribute to the further development of brain research. The extent to which we achieve this goal will be determined by the utility of these books.

Sidney A. Simon, Ph.D.
Miguel A.L. Nicolelis, M.D., Ph.D.
Series Editors

The Editors

Cynthia Kuhn, Ph.D., is a professor in the Department of Pharmacology and Cancer Biology at Duke University Medical Center. She earned her B.A. in biology at Stanford University and her Ph.D. in pharmacology at Duke. She has been a faculty member at Duke since 1978. She has studied the effect of addictive drugs on the developing brain, and the interaction of hormonal state and the effects of addictive drugs throughout her career. Dr. Kuhn's work has identified the contribution of enhanced dopaminergic function to the addiction risk for adolescents and females. She has received continuous funding from the National Institute of Drug Abuse to study the effects of cocaine on women and adolescent animals and has published more than 250 scientific papers. She has trained 7 predoctoral, 10 postdoctoral, and more than 50 undergraduate students.

In addition to her research activities, Dr. Kuhn is extremely active in teaching for both professional and lay audiences. She has taught pharmacology (drug abuse, endocrinology) in the medical school curriculum for 25 years and teaches an undergraduate course at Duke entitled "Drugs and the Brain." She is a member of the Translational Prevention Research Center at Duke and is active nationally in drug abuse education for students, teachers, prevention specialists and treatment providers. She has coauthored three books for the lay public: *Buzzed: The Straight Facts About the Most Used and Abused Drugs From Alcohol to Ecstasy* (Norton, 2008), *Just Say Know: Talking with Kids about Drugs and Alcohol* (Norton, 2002), and *Pumped: Straight Facts for Athletes About Drugs, Supplements and Training* (Norton, 2000). She lectures to lay audiences including parents, students, church groups and others.

George F. Koob, Ph.D., is a professor and chair of the Committee on the Neurobiology of Addictive Disorders at The Scripps Research Institute, adjunct professor in the Departments of Psychology and Psychiatry, and adjunct professor in the Skaggs School of Pharmacy and Pharmaceutical Sciences at the University of California, San Diego. Dr. Koob received his bachelor of science degree from Pennsylvania State University and his Ph.D. in Behavioral Physiology from The Johns Hopkins University. An authority on addiction and stress, Dr. Koob has published over 670 scientific papers and has received continuous funding for his research from the National Institutes of Health, including the National Institute on Alcohol Abuse and Alcoholism (NIAAA) and the National Institute on Drug Abuse (NIDA). He is director of the NIAAA Alcohol Research Center at The Scripps Research Institute, consortium coordinator for NIAAA's multi-center Integrative Neuroscience Initiative on Alcoholism, and co-director of the Pearson Center for Alcoholism and Addiction Research. He has trained 10 predoctoral and 64 postdoctoral fellows. Dr. Koob is editor-in-chief USA for the journal *Pharmacology Biochemistry and Behavior* and editor-in-chief for the *Journal of Addiction Medicine*. He won the Daniel Efron Award for excellence in research from the American College of Neuropsychopharmacology, was honored as a Highly Cited Researcher from the Institute for Scientific Information, was presented with the Distinguished Investigator Award from the Research Society on Alcoholism, and won the Mark Keller Award from NIAAA. He published a landmark book in 2006 with his colleague Dr. Michel Le Moal, *Neurobiology of Addiction* (Academic Press-Elsevier, Amsterdam).

Dr. Koob's research interests have been directed at the neurobiology of emotion, with a focus on the theoretical constructs of reward and stress. He has made contributions to our understanding of the anatomical connections of the emotional systems and the neurochemistry of emotional function. Dr. Koob has identified afferent and efferent connections of the basal forebrain (extended amygdala) in the region of the nucleus accumbens, bed nucleus of the stria terminalis, and central nucleus of the amygdala in motor activation, reinforcement mechanisms, behavioral responses to stress, drug self-administration, and the neuroadaptation associated with drug dependence.

Dr. Koob is one of the world's authorities on the neurobiology of drug addiction. He has contributed to our understanding of the neurocircuitry associated with the acute reinforcing effects of drugs of abuse and more recently on the neuroadaptations of these reward circuits associated with the transition to dependence. He has validated key animal models for dependence associated with drugs of abuse and has begun to explore a key role of anti-reward systems in the development of dependence.

Dr. Koob's work with the neurobiology of stress includes the characterization of behavioral functions in the central nervous system for catecholamines, opioid peptides, and corticotropin-releasing factor. Corticotropin-releasing factor, in addition to its classical hormonal functions in the hypothalamic-pituitary-adrenal axis, is also located in extrahypothalamic brain structures and may have an important role in brain emotional function. Recent use of specific corticotropin-releasing factor antagonists suggests that endogenous brain corticotropin-releasing factor may be involved in specific behavioral responses to stress, the psychopathology of anxiety and affective disorders, and drug addiction. Dr. Koob also has characterized functional roles for other stress-related neurotransmitters/neuroregulators such as norepinephrine, vasopressin, hypocretin (orexin), neuropeptide Y, and neuroactive steroids.

The identification of specific neurochemical systems within the basal forebrain system of the extended amygdala involved in motivation has significant theoretical and heuristic impact. From a theoretical perspective, identification of a role for dopaminergic, opioidergic, GABAergic, glutamatergic, and corticotropin-releasing factor systems in excessive drug taking provides a neuropharmacologic basis for the allostatic changes hypothesized to drive the process of pathology associated with addiction, anxiety, and depression. From a heuristic perspective, these findings provide a framework for further molecular, cellular, and neurocircuit research that will identify the basis for individual differences in vulnerability to pathology.

Contributors

Anne Beck
Department of Psychiatry und
  Psychotherapy, Charité
Universitätsmedizin Berlin
Campus Charité Mitte
Berlin, Germany

Paul A. Garris
Department of Biological Sciences
and
Department of Chemistry
Illinois State University
Normal, Illinois

Anthony A. Grace
Departments of Neuroscience,
  Psychiatry, and Psychology
University of Pittsburgh
Pittsburgh, Pennsylvania

Danielle L. Graham
CNS Pharmacology
Merck Research Labs
Boston, Massachusetts

Sabine M. Grüsser
Institute of Medical Psychology, Charité
Universitätsmedizin Berlin
Campus Charité Mitte
Berlin, Germany

Karine Guillem
Department of Psychiatry
University of Pennsylvania School of
  Medicine
Philadelphia, Pennsylvania

Benjamin Y. Hayden
Department of Neurobiology
Center for Neuroeconomic Studies
and
Center for Cognitive Neuroscience
Duke University Medical Center
Durham, North Carolina

Andreas Heinz
Department of Psychiatry und
  Psychotherapy, Charité
Universitätsmedizin Berlin
Campus Charité Mitte
Berlin, Germany

Jeffrey T. Klein
Department of Neurobiology
Center for Neuroeconomic Studies
and
Center for Cognitive Neuroscience
Duke University Medical Center
Durham, North Carolina

George F. Koob
Committee on the Neurobiology of
  Addictive Disorders
The Scripps Research Institute
La Jolla, California

Alexai V. Kravitz
Department of Psychiatry
and
Neuroscience Graduate Group
University of Pennsylvania School of
  Medicine
Philadelphia, Pennsylvania

Cynthia M. Kuhn
Department of Pharmacology and
  Cancer Biology
Duke University Medical Center
Durham, North Carolina

Jan Mir
Department of Psychiatry und
  Psychotherapy, Charité
Universitätsmedizin Berlin
Campus Charité Mitte
Berlin, Germany

Laura L. Peoples
Department of Psychiatry
and
Neuroscience Graduate Group
University of Pennsylvania School of
  Medicine
Philadelphia, Pennsylvania

Michael L. Platt
Department of Neurobiology
Center for Neuroeconomic Studies
and
Center for Cognitive Neuroscience
Duke University Medical Center
Durham, North Carolina

Stefan G. Sandberg
Department of Biological Sciences
Illinois State University
Normal, Illinois

David W. Self
Department of Psychiatry
Seay Center for Basic and Applied
  Research in Psychiatric Illness
University of Texas Southwestern
  Medical Center
Dallas, Texas

Stephen V. Shepherd
Neuroscience Institute
Princeton University
Princeton, New Jersey

Karli K. Watson
Department of Neurobiology
Center for Neuroeconomic Studies
and
Center for Cognitive Neuroscience
Duke University Medical Center
Durham, North Carolina

Friedbert Weiss
Molecular and Integrative Neurosciences
  Department (SP30-2120)
Scripps Research Institute
La Jolla, California

Jana Wrase
Department of Psychiatry und
  Psychotherapy, Charité
Universitätsmedizin Berlin
Campus Charité Mitte
Berlin, Germany

1 Advances in Animal Models of Relapse for Addiction Research

Friedbert Weiss

CONTENTS

1.1 INTRODUCTION AND SCOPE

1.1.1 THE NEUROBEHAVIORAL BASIS OF DRUG SEEKING AND RELAPSE

Drug addiction is a chronically relapsing disorder characterized by compulsive drug-seeking and use (Leshner 1997; McLellan et al. 2000; O'Brien et al. 1998; O'Brien and McLellan 1996). Long-lasting vulnerability to relapse has been recognized as a phenomenon pivotal for the understanding and treatment of drug addiction. Elucidation of the neurobiological mechanisms underlying the chronically relapsing nature of addiction and identification of pharmacological treatment targets for relapse prevention has therefore emerged as the central issue of importance in addiction research. The clinical as well as experimental literature implicates three major factors precipitating craving and relapse. One of these is learned responses evoked by environmental stimuli that have become associated with the subjective actions

1

of drugs of abuse by means of classical conditioning. Exposure to such stimuli can evoke drug desire and drug seeking, effects that have been implicated both in maintaining ongoing drug use and eliciting drug-seeking during abstinence (Everitt et al. 2001; Littleton 2000; O'Brien et al. 1998; See 2002; Van De Laar et al. 2004). Drug-related stimuli can also elicit automatic responses that lead to drug-seeking and relapse without recruiting conscious desire or distinct feelings of craving (Ingjaldsson et al. 2003; Miller and Gold 1994; Stormark et al. 1995; Tiffany and Carter 1998; Tiffany and Conklin 2000). Studies in animals have confirmed that environmental stimuli that have become associated with the reinforcing actions of drugs of abuse reliably elicit drug-seeking in animal models of relapse drug-seeking behavior (for review, see See 2002; Shaham and Miczek 2003; Shaham et al. 2003; Shalev et al. 2002). In particular, contextual stimuli predictive of cocaine (Ciccocioppo et al. 2004b, 2001b; Weiss et al. 2000, 2001), ethanol (Ciccocioppo et al. 2001a, 2003, 2002; Katner et al. 1999; Katner and Weiss 1999; Liu and Weiss 2002b), or heroin (Gracy et al. 2000) availability reliably elicit strong recovery of extinguished drug-seeking behavior. Indeed, drug seeking induced by these stimuli shows remarkable resistance to extinction (Weiss et al. 2001; Ciccocioppo et al. 2001a) and, in the case of cocaine, can still be observed after several months of abstinence (Ciccocioppo et al. 2001b; Weiss et al. 2001). Moreover, cocaine seeking induced by drug-related stimuli can progressively increase in strength (incubate) during long-term abstinence (Grimm et al. 2001). Overall, the persistence of the behavioral effects of drug-associated environmental cues resembles the persistence of conditioned cue reactivity and cue-induced craving in humans (e.g., Childress et al. 1993).

Stress is the second factor with an established role in relapse to drug and alcohol use in humans (e.g., Brown et al. 1995; Kreek and Koob 1998; Marlatt 1985; McKay et al. 1995; Sinha 2000, 2001; Sinha et al. 1999, 2000, 2006). The significance of stress in drug seeking and use is also well documented in the animal literature. Various stressors facilitate acquisition of drug self-administration or increased drug intake in rodents (Blanchard et al. 1987; Goeders and Guerin 1994; Haney et al. 1995; Higley et al. 1991; Mollenauer et al. 1993; Nash and Maickel 1988; Ramsey and Van Ree 1993). More importantly, with respect to the present discussion, stress consistently elicits reinstatement of ethanol seeking in drug-free animals, with footshock being the predominant model of stress (for review, see Le and Shaham 2002; Sarnyai et al. 2001; Shaham et al. 2000; Shalev et al. 2002).

A third factor with a major role in relapse is neuroadaptive dysregulation induced by chronic drug use. Such disturbances are thought to underlie symptoms of anxiety, irritability, autonomic arousal, and exaggerated responsiveness to anxiogenic stimuli that emerge when drug use is discontinued (Kajdasz et al. 1999; Kampman et al. 2001; McDougle et al. 1994). Growing evidence suggests that neuroadaptive changes outlast physical withdrawal and detoxification (Kowatch et al. 1992). For example, detoxified cocaine addicts exhibit increased panic and anxiety (Goodwin et al. 2002; Razzouk et al. 2000; Rounsaville et al. 1991; Walfish et al. 1990; Ziedonis et al. 1994) that may result directly from prior cocaine use (Aronson and Craig 1986; Blanchard and Blanchard 1999). Anxiety and other symptoms, such as drug craving, sleep dysregulation, and somatic symptoms, predict poor outcome (Kasarabada et al. 1998). Such "protracted withdrawal" symptoms, originally described in opiate addicts (Martin and Jasinski 1969), also represent a common complication in patients recovering from cocaine and alcohol addiction (Angres and Benson 1985; Gawin and Kleber 1986; Kreek 1987; Meyer 1996; Satel et al. 1993) and introduce alleviation of discomfort and negative affect and, thus, *negative reinforcement* as a motivational basis for relapse.

This chapter reviews animal models of relapse with emphasis on recent applications to study and elucidate (a) the role of dependence history in susceptibility to relapse, (b) interactions among risk factors for relapse in eliciting drug-seeking behavior, and (c) procedures designed to study and differentiate the distinctly compulsive nature of drug-seeking as opposed to behavior motivated by natural rewards. This review will center on alcohol-seeking behavior as a model. However, the issues to be addressed apply to other drugs of abuse as well, and where applicable, the reader is referred to the appropriate literature.

1.1.2 MODELING RELAPSE IN ANIMALS

The most widely employed animal model of craving and relapse use is the extinction-reinstatement model. This model has perhaps been most extensively applied for investigations of the significance of environmental stimuli conditioned to the reinforcing actions of drugs of abuse in the relapse process, but is also widely employed to study the effects of stress and drug priming on the resumption of drug seeking. Here, we will focus predominantly on reinstatement in the context of conditioning studies.

The model, its range of applications, and its limitations have been reviewed extensively elsewhere (See 2005; Shaham et al. 2003; Shalev et al. 2002; Katz and Higgins 2003; Shaham and Miczek 2003; Weiss 2005). Briefly, several variants of the model exist. In the most commonly employed procedure, animals are trained to respond at an operandum (e.g., lever or nose-poke sensor), and completion of a given schedule requirement results in delivery of the drug. Each drug-reinforced response is paired with brief presentation of one or more environmental stimuli (e.g., a tone, cue light). In this variant of the model, both drug administration and presentation of these conditioned stimuli (CS) are contingent upon the animal's operant response. Use of a compound stimulus (i.e., concurrent presentation of two or more CS) typically produces more robust conditioned reinstatement than a single stimulus (See et al. 1999). Once reliable drug self-administration is acquired, drug-reinforced responding is extinguished by withholding both the drug and CS until a given extinction criterion is reached. Reinstatement tests then are conducted in which the degree of recovery of responding at the previously drug-paired operandum, now maintained by response-contingent presentation of the CS only, serves as a measure of craving or relapse. Extinction and reinstatement tests can be conducted according to a *within-session*, *between-session*, and *within-between session* sequence. In the *within-session* procedure, a single extinction session is conducted, followed immediately by the reinstatement test (Figure 1.1A). In the *between-session* protocol, extinction sessions are conducted daily, and reinstatement tests commence typically one day after the extinction criterion is reached (Figure 1.1C). This procedure "adds" a drug-free (i.e., abstinence) period to the mere extinction of drug-reinforced responding. In the

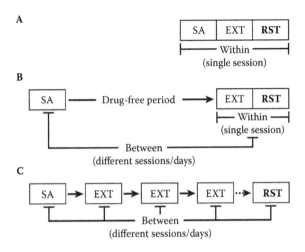

FIGURE 1.1 Illustration of the temporal sequence of drug self-administration (SA), extinction (EXT), and reinstatement test (RST) sessions in three variants of the extinction-reinstatement model: the within-session (A), between-within (B), and between-session (C) procedure.

between-within procedure, no extinction sessions are conducted after rats acquire drug-reinforced responding. Instead, an abstinence period is imposed, after which extinction and reinstatement sessions are conducted in a within-session manner (Figure 1.1B). In this application of the model, extinction responses (i.e., resistance to extinction) are often taken as a measure of craving and drug seeking, rather than reinstatement responses (Grimm et al. 2001; Lu et al. 2004). The great majority of contemporary applications of the extinction-reinstatement model employ the between-session or between-within-session procedure.

A second variant of the extinction-reinstatement procedure is the contextual model. The procedures here are similar to those above, except that contextual stimuli present continuously throughout the drug self-administration are used for conditioning. That is, these stimuli are neither discretely paired with drug delivery nor contingent upon a response. Multiple contextual reinstatement procedures exist. One utilizes differential reinforcement of behavior in the presence of *discriminative stimuli*. In this procedure, during self-administration learning, responses at the operandum are reinforced by the drug only in the presence of this stimulus. In the absence of this stimulus (or the presence of an alternative stimulus), responses remain nonreinforced. Owing to their predictive nature for drug availability, discriminative stimuli "set the occasion" for engaging in responding at the drug-paired lever (i.e., lead to the initiation of responding). Additionally, by virtue of their presence during drug consumption, these stimuli also become associated with the rewarding effects of the drug and thus acquire incentive motivational valence. As a result, these stimuli are particularly powerful in eliciting drug seeking and reinstatement. Another frequently employed contextual model, based on procedures by Bouton and colleagues (Bouton and Bolles 1979; Bouton and Swartzentruber 1986), utilizes distinct environments that provide compound contextual cues (i.e., concurrent presence of olfactory, auditory, tactile, and visual cues) to produce *renewal* of extinguished reward seeking. Briefly, in this model, responding is reinforced by a given (drug) reinforcer in one context. Instrumental responding then is extinguished in a second context. Subjects subsequently tested in the second context show low drug seeking because the behavior has been extinguished in this context. In contrast, drug seeking shows reactivation or renewal (i.e., nonreinforced responding at the previously active operandum) in animals tested in the first (drug-paired) context.

1.1.3 Significance of Considering Dependence History in Animal Models of Relapse

In the case of alcohol addiction, ample evidence exists to show that alcohol-associated stimuli or events can evoke drug desire that may lead to the resumption of drinking in abstinent alcoholics (Cooney et al. 1987, 1997; Eriksen and Gotestam 1984; Kaplan et al. 1985; Laberg 1986; Monti et al. 1987, 1993).

The animal literature confirms a significant role for such conditioning factors in alcohol seeking and relapse. Studies using the extinction-reinstatement model show that stimuli discretely paired with alcohol delivery and alcohol-related contextual stimuli exert powerful and long-lasting control over ethanol-seeking behavior. Such stimuli reliably reinstate extinguished ethanol-seeking behavior as measured by the resumption of responding at a previously active ethanol-paired lever (Bachteler et al. 2005; Bienkowski et al. 1999; Ciccocioppo et al. 2001a, 2004a, 2003, 2002; Katner et al. 1999; Katner and Weiss 1999; Liu and Weiss 2002a, 2002b). The response-reinstating effects of ethanol-related contextual stimuli are surprisingly resistant to extinction in that the efficacy of these cues to elicit resumption of ethanol-seeking behavior does not diminish even when presented repeatedly under non-reinforced conditions (Figure 1.2A,B)—in contrast to behavior induced by stimuli conditioned to highly palatable natural reward (Figure 1.3). In this context, it should be noted that the data in Figures 1.2 and 1.3 document that the initial efficacy of reward-predictive contextual stimuli to elicit behavior directed at obtaining a given reward is similar regardless of whether the behavior is directed at obtaining hedonic gain associated with drug or alcohol reward versus nondrug reward. However, the pattern of differential resistance to extinction with repeated exposure suggests that the conditioned incentive effects of alcohol or drug-related stimuli do not resemble the effects of stimuli conditioned to even highly desirable conventional reinforcers. Specifically drug-related environmental stimuli exert longer-lasting control over behavior than stimuli conditioned to potent conventional reinforcers. This therefore suggests that learning and associative processes responsible for ethanol-related conditioning are distinct from those mediating associations between stimuli paired with the rewarding effects of conventional reinforcers and, presumably, these specific drug- or ethanol-related associative processes lead to compulsive-like drug seeking. Consistent with clinical findings, reinstatement induced by alcohol cues is sensitive to reversal by opioid antagonist administration (Bienkowski et al. 1999; Burattini et al. 2006; Ciccocioppo et al. 2003, 2002; Katner et al. 1999). In alcoholics, naltrexone attenuates cue-induced craving (Monti et al. 1999; Rohsenow et al. 2000) and reduces

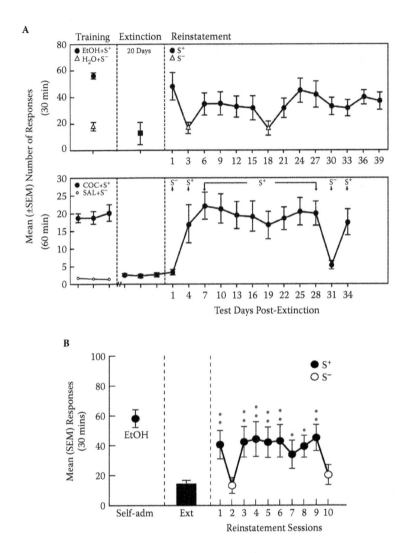

FIGURE 1.2 (A) Persistence of the effects of a drug-related discriminative stimulus (S$^+$) determined at 3-day intervals over a 39-day period for ethanol (EtOH) in Indiana Alcohol Preferring P rats (upper panel; modified from Ciccocioppo R, Angeletti S, Weiss F (2001a) Long-lasting resistance to extinction of response reinstatement induced by ethanol-related stimuli: role of genetic ethanol preference. *Alcohol Clin Exp Res* 25: 1414–1419) and a 34-day period for cocaine (lower panel; modified from Weiss F, Martin-Fardon R, Ciccocioppo R, Kerr TM, Smith DL, Ben-Shahar O (2001) Enduring resistance to extinction of cocaine-seeking behavior induced by drug-related cues. *Neuropsychopharmacology* 25: 361–372). Note that responses remained at (or returned to) extinction levels when rats were presented with a discriminative stimulus conditioned to absence of reward (S$^-$: days 1 and 31, top; days 3 and 18, bottom). All S$^+$ effects were significant at the $p < 0.01$ (cocaine) or $p < 0.05$ (ethanol) level. (B) Persistence of the effects of an EtOH-predictive S$^+$ over multiple reinstatement tests in genetically heterogeneous Wistar rats conducted every third day for a total of 10 sessions. Note that responding returned to extinction levels when an S$^-$ was presented (Sessions 2 and 10). $**p < 0.01$, vs. extinction (Ext).

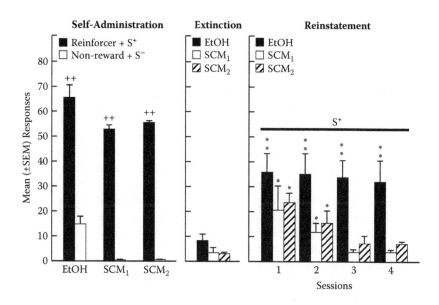

FIGURE 1.3 Self-administration: responding reinforced by 10% (w/v) ethanol (EtOH) or sweetened condensed milk (SCM; representing original results (SCM_1) and a replication (SCM_2)) in the presence of a discriminative stimulus (S^+) associated with the availability of EtOH or SCM. Data represent averages across the final three sessions of the self-administration/conditioning phase. Extinction: Non-reinforced responses averaged across the final three extinction sessions. Reinstatement: Responses during reinstatement tests conducted every third day for a total of four tests in the presence of the S^+ previously predictive of EtOH or SCM reward. EtOH and SCM were not available during the reinstatement test phase. $**p < 0.01$; $*p < 0.05$, different from Extinction.

relapse rates (O'Brien et al. 1996; Volpicelli et al. 1992). Moreover, excellent correspondence exists between neural mapping studies in animals (Dayas et al. 2007; Zhao et al. 2006) and functional brain imaging studies in drinkers (e.g., Braus et al. 2001; George et al. 2001; Kareken et al. 2004; Myrick et al. 2004; Schneider et al. 2001) with respect to the neurocircuitry activated by alcohol cue manipulations that, in humans, result in self-reports of craving and, in animals, alcohol-seeking behavior. Conditioned reinstatement of ethanol seeking in animals, therefore, has good predictive validity as a model of craving and relapse linked to alcohol cue exposure.

The literature on alcohol cue effects in animal models of relapse complements the clinical literature on conditioned cue reactivity and craving in alcoholics and confirms that alcohol cue conditioning is a critical factor in the susceptibility to relapse. What must be kept in mind, however, is that animal studies are typically conducted in rats with a history of restricted daily access (30–60 min) to ethanol, during which typically only modest blood alcohol levels of 50–80 mg% are attained. Conditioned reinstatement induced by alcohol cues under these conditions is consistent with evidence that even light drinkers that are not dependent show conditioned cue reactivity and mild craving in response to alcohol cue exposure (Greeley et al. 1993; Streeter et al. 2002). However, cue-induced ethanol seeking in nondependent animals depends on conditioning of alcohol cues to the positive reinforcing effects of modest ethanol doses only, and is therefore limited with respect to modeling the learning history and motivational forces underlying conditioned cue reactivity and craving in alcoholics with long histories of heavy drinking and repeated withdrawal episodes.

Cessation of ethanol intake after chronic use leads to symptoms of negative affect including, among others, anxiety, mood disturbances, and exaggerated responsiveness to stress. Data from animal models have confirmed that major consequences of *acute* withdrawal from chronic ethanol exposure are increased sensitivity to anxiogenic stimuli as measured in the elevated plus maze, acoustic startle, and social interaction tests (Breese et al. 2004; Knapp et al. 2004; Overstreet et al. 2004; Rassnick et al. 1993) as well as in reward deficits (Schulteis et al. 1995). These findings suggest that symptoms of negative affect that result from ethanol withdrawal motivate the maintenance and resumption of ethanol intake. Consistent with this hypothesis, pharmacological manipulations that reverse stresslike and anxiogenic consequences of ethanol withdrawal (Macey et al. 1996; Valdez et al. 2002) also reverse withdrawal-associated increases in ethanol self-administration (Roberts et al. 1996; Valdez et al. 2002, 2003). Clinical studies support the hypothesis that negative affect symptoms during abstinence maintain drinking and promote relapse through negative reinforcement mechanisms (e.g., Cooney et al. 1997; Hershon 1977; Sinha et al. 2000). Most importantly, in alcoholics the severity of craving induced by alcohol cues is highly correlated with the history and severity of dependence (Greeley et al. 1993; Laberg 1986; Streeter et al. 2002). A likely explanation for these findings is that consumption of ethanol during withdrawal allows for new learning to occur (i.e., the reversal or avoidance of withdrawal distress by resumption of ethanol use), thereby modifying an individual's ethanol reinforcement history to include learning about negative reinforcement as an important aspect of ethanol's actions. As a result, ethanol becomes a qualitatively different and more potent reinforcer. Clearly, environmental stimuli can become associated with both the positive and negative aspects of ethanol's reinforcing actions. However, the significance of a history of negative reinforcement by ethanol for craving and relapse associated with ethanol cue exposure has not been extensively studied in animal models.

With these considerations in mind, efforts are beginning to be made to examine and model the significance of a history of dependence in the relapse process. The following sections will examine the application and potential of the extinction-reinstatement model to elucidate the role of dependence history in susceptibility to relapse as well as interactions among relapse risk facts in producing craving and relapse, drawing from studies in alcohol reinforcement and addiction literature.

1.2 MODELING THE INCENTIVE EFFECTS OF DRUG CUES IN DEPENDENT SUBJECTS

Given the motivational significance of a dependence history discussed above, a hypothesis testable in animal models is that self-administration of ethanol during withdrawal increases the incentive salience of ethanol-related environmental stimuli. Specifically, two questions can be asked: (1) does learning about escape from the aversive effects of withdrawal by active ethanol self-administration alter alcohol-seeking induced by stimuli that had previously been associated only with the positive reinforcing effects of ethanol, and (2) to what extent do stimuli that are associated specifically with ethanol self-administration during withdrawal (i.e., negative reinforcement) promote reinstatement of ethanol-seeking behavior?

1.2.1 SIGNIFICANCE OF HISTORY OF NEGATIVE REINFORCEMENT IN REINSTATEMENT

In a study addressing the first question, ethanol reinforced responses in nondependent rats were response-contingently paired with a discrete light cue. Rats then were either made dependent by ethanol vapor intoxication or remained nondependent. During the last 3 days of dependence induction, rats were removed daily from the ethanol vapor chambers and allowed to operantly self-administer ethanol for 12 h, followed by reexposure to ethanol vapor for 12 h. The rats then were fully withdrawn from ethanol, and ethanol-reinforced operant responding was extinguished in daily sessions over a 3-week period. Animals then were tested for reinstatement by response-contingent presentation of the ethanol cue that had been conditioned to the positive reinforcing effects of ethanol during self-administration training. As shown in Figure 1.4 (EXT vs. EtOH CS), ethanol-dependent rats that had been given the opportunity to self-administer ethanol during withdrawal showed significantly greater conditioned reinstatement than nondependent rats that had been given the same amount of exposure to the cue than the dependent animals (Liu and Weiss 2002a), indicating that learning about negative ethanol reinforcement during withdrawal enhances drug-seeking by cues previously associated only with positive reinforcement by ethanol. One should note that, procedurally, these findings also exemplify differences in the robustness of conditioning associated with a simple versus compound CS. Ethanol reinforcement in this study had been conditioned to a simple CS (light cue) only, rather than a compound stimulus (See et al. 1999). As alluded to above, the former class of stimuli are typically less effective in reinstating extinguished drug-seeking behavior than compound stimuli. Thus, in contrast to studies employing compound stimuli or contextual conditioning procedures, only subthreshold effects on reinstatement were obtained with the simple CS in nondependent rats.

1.2.2 SIGNIFICANCE OF STIMULI CONDITIONED TO NEGATIVE REINFORCEMENT IN REINSTATEMENT

Data also exist pertinent to the second question, namely, whether stimuli conditioned specifically to the withdrawal-ameliorating actions of ethanol acquire motivational significance in terms of eliciting alcohol-seeking behavior. Here, the same conditioning and reinstatement procedures as above were used, except that a second and distinctly different CS was introduced, paired specifically and only with ethanol-reinforced responses during the 12-h self-administration sessions conducted over the 3-day withdrawal phase. The effects of this stimulus on reinstatement then were tested during acute ethanol withdrawal and 3 weeks following termination of ethanol vapor exposure. Both the CS that had been paired with the positive reinforcing

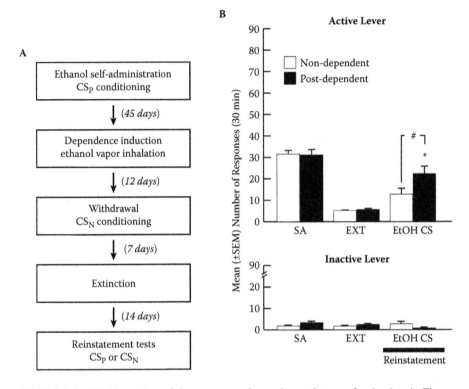

FIGURE 1.4 (A) Illustration of the sequence of experimental stages for the data in Figure 1.4B. (B) Responses at a previously active and inactive lever in ethanol nondependent and dependent rats trained to operantly self-administer ethanol, with presentation of an EtOH CS response contingently paired with delivery of each EtOH (0.1 ml) reinforcer. SA: EtOH-reinforced responses at the end of self-administration training. EXT: Responses during the final three extinction sessions. Reinstatement: Responses during response-contingent presentation of an ethanol-associated conditioned stimulus (EtOH CS). *$p < 0.05$, **$p < 0.01$, different from extinction. #$p < 0.05$ differences between nondependent and postdependent rats (modified from Liu and Weiss 2002, *J. Neurosci* 22: 7856–7861, with permission).

actions of ethanol prior to induction of dependence only and the stimulus that had been paired with negative reinforcement during withdrawal only produced substantial reinstatement during acute withdrawal (Figure 1.5A). Robust reinstatement was still obtained in rats tested following 3 weeks of abstinence (Figure 1.5B,C). These observations reveal that stimuli conditioned specifically to amelioration of withdrawal are of substantial motivational significance and elicit or contribute to ethanol-seeking behavior in their own right.

With respect to animal model development, an important point to be taken from findings such as these is that an induction of alcohol dependence alone, without providing an opportunity for negative reinforcement by ethanol to occur, is not sufficient to document and study the significance of ethanol cues associated with a history of dependence. This point is exemplified by an earlier study that compared reinstatement produced by a contextual ethanol cue in nondependent rats to the effects of

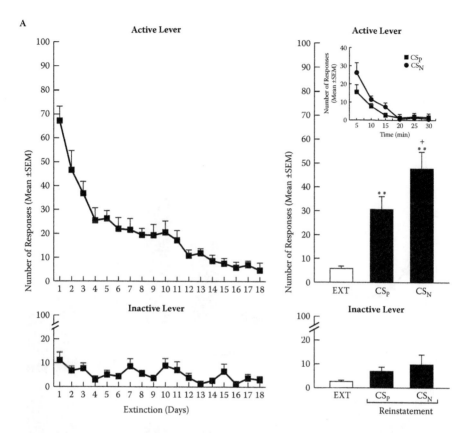

FIGURE 1.5 (A) Cue-induced response reinstatement during acute ethanol withdrawal. The left panels show responses at the active and inactive lever during the extinction phase preceding reexposure of the animals to ethanol vapor and the subsequent reinstatement test. The right panels show responses averaged across the last 3 days of the extinction phase (EXT), and lever-presses associated with response-contingent presentation of the stimulus previously paired with ethanol in the nondependent state (CS_P) or during withdrawal (CS_N). Inset: Time course of responding in the CS_P and CS_N test conditions. **$p < 0.01$, different from EXT; $+ p < 0.05$, different from CS_P.

the same cue in rats with a history of dependence that consisted of either single or multiple ethanol vapor intoxication and withdrawal (Ciccocioppo et al. 2003). In this study, ethanol was initially self-administered in 30-min sessions, and the availability of ethanol was conditioned to the presence of a discriminative stimulus (S^D). The rats then were subjected to single or multiple ethanol vapor intoxication, followed by 7 days of full withdrawal from ethanol. Responding at the lever previously producing ethanol during the self-administration of the experiment then was extinguished, and rats were tested for reinstatement in the presence of the ethanol-associated S^D only. Contrary to expectations, the effects of this ethanol cue on the number of reinstatement responses did not differ as a function of either a single or repeated withdrawal history from that in nondependent rats. Examination of the contingencies that were in effect in this study reveals that two conditions had not been met that would be

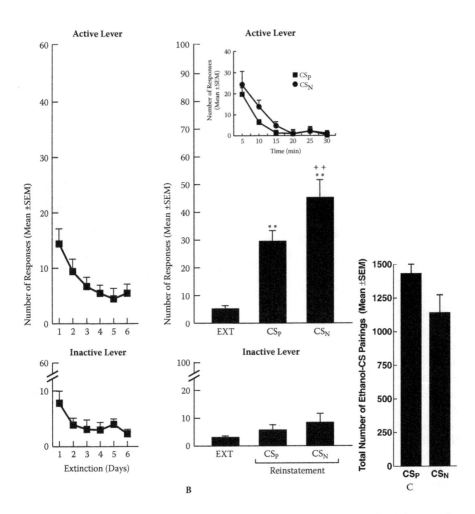

FIGURE 1.5 (continued) (B) Cue-induced response reinstatement determined three weeks following withdrawal. The left panels show responses at the active and inactive lever during the set of extinction sessions preceding this second reinstatement test. The center panels show the mean number of responses during the last three extinction sessions (EXT), and lever-presses associated with response-contingent presentation of the stimulus previously paired with ethanol in the nondependent state (CS_P) or during withdrawal (CS_N). Inset: Time course of responding in the CS_P and CS_N test conditions. **$p < 0.01$, different from EXT; ++$p < 0.01$, different from CS_P. (C). The right panel shows the mean (± SEM) total number of pairings of ethanol-reinforced responses with the CS_P during the conditioning phase in the nondependent state and with the CS_N during the three 12-hour conditioning sessions in the withdrawal state (for animals shown in Figure 1.2B). No differences were found in the total number of pairings between ethanol delivery (following completion of each FR 3 ratio requirement) and the respective conditioned stimuli ($t_{11} = 1.92$; NS in ethanol nondependent (CS_P) and dependent (CS_N) rats). A total of 1434 ± 68.7 discrete pairings occurred during the 45-day CS_P conditioning phase, and a total of 1142 ± 131.7 pairings during the three 12-hour CS_N conditioning session.

required for a valid test of the hypothesis that a history of negative reinforcement by ethanol modifies or enhances the motivating effects of alcohol-related environmental stimuli. First, the ethanol S^D used to test for conditioned reinstatement had been paired with ethanol availability only while rats were in the nondependent state and, therefore, was associated only with the acute, positive reinforcing effects of ethanol. Second, although in rats subjected to multiple withdrawal cycles, reexposure to ethanol vapor did provide an opportunity for associations to develop between ethanol vapor and alleviation of withdrawal, these animals had been subjected to ethanol withdrawal without being given the opportunity to associate an active behavioral response (i.e., oral ethanol self-administration) with reversal of the aversive consequences of withdrawal.

1.3 MODELING INTERACTIONS AMONG RISK FACTORS FOR RELAPSE

Abstinent drug addicts likely are frequently exposed to multiple external risk factors, while at the same time suffering various degrees of anxiety or mood-dysregulation resulting from neuroadaptive changes. Consideration of the concurrent operation of multiple risk factors will be of importance in animal model development, both to gain a better understanding of the relapse process per se and to improve the likelihood of identifying effective treatment targets. For example, whereas both craving and stress have been implicated as critical risk factors for relapse, the predictive relationship between subjective reports of stress or craving and subsequent actual relapse remains controversial (Drummond 2001; Katz and Higgins 2003; Sinha 2001; Tiffany et al. 2000). However, risk factors for relapse are typically studied in isolation, that is, conditions that exclude consideration of the impact of multiple external risk factors. Indeed, it seems likely that the probability of relapse varies as a function of the number and intensity of risk factors operative at a given time, with relapse occurring when the sum of these motivating forces reaches a critical threshold. Several recent studies have begun to address this issue in reinstatement models.

1.3.1 INTERACTIVE EFFECTS OF DRUG CUES AND STRESS

Stress has an established role in alcohol abuse in humans and is a major determinant of relapse (Brown et al. 1995; Marlatt 1985; McKay et al. 1995; Sinha 2000, 2001; Sinha et al. 2003; Wallace 1989). The significance of stress in drug-seeking behavior is well documented also in the animal literature. Physical, social, and emotional stress can facilitate acquisition or increase self-administration of ethanol in rodents and nonhuman primates (Blanchard et al. 1987; Higley et al. 1991; Mollenauer et al. 1993; Nash and Maickel 1988).

Importantly with respect to the present discussion, stress consistently elicits reinstatement of ethanol-seeking in drug-free animals, with footshock being the predominant model of stress (see Le and Shaham 2002 for review). Interactive effects of stress and drug-related cues on drug seeking were modeled in an original series of studies by testing for the concurrent effects of footshock stress and an ethanol-

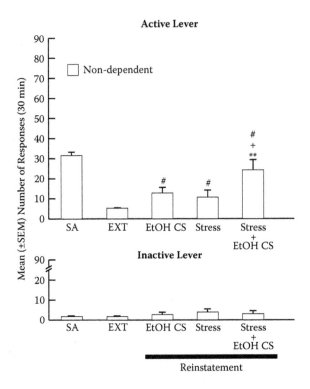

FIGURE 1.6 Responses at a previously active and inactive lever in nondependent rats trained to operantly self-administer ethanol. SA: Ethanol-reinforced responses at the end of self-administration training. EXT: Responses during the final three extinction sessions. Reinstatement: Responses after intermittent footshock (Stress), during response-contingent presentation of an ethanol-associated conditioned stimulus (EtOH CS), and response-contingent presentation of the EtOH CS following footshock (Stress + EtOH CS). $*p < 0.05$, $**p < 0.01$, different from extinction. $+p < 0.05$, $++p < 0.01$, different from Stress and EtOH CS (modified from Liu and Weiss, 2002, *J Neurosci* 22: 7856–7861, with permission).

associated CS on reinstatement (Liu and Weiss 2002a, 2003). Rats were trained to self-administer ethanol, and reinforced responses were paired with brief presentation of a simple CS. After withdrawal, ethanol-reinforced responding was extinguished, and the reinstatement of alcohol-seeking behavior was studied under three conditions: (1) during response-contingent presentation of the CS alone, (2) after exposure to 10 min of intermittent footshock stress alone, and (3) during response-contingent presentation of the CS following exposure to footshock stress. Under these conditions the ethanol CS and footshock, when presented alone, produced only threshold effects on alcohol-seeking behavior. However, the ethanol CS elicited strong responding in animals that had been subjected to footshock stress before the session (Figure 1.6; EtOH CS and Stress). Thus, interactive effects between stress and drug-related cues exacerbating drug-seeking can readily be demonstrated in animal models. Indeed, the existence of such interactive effects has recently been confirmed with another drug of abuse and in the context of a different stress manipulation.

Yohimbine—a noradrenergic α_2 receptor antagonist with anxiogenic action—that has found increasing application as an alternative stressor in animal models of drug seeking (Feltenstein and See 2006; Marinelli et al. 2007) and reward seeking (Ghitza et al. 2006; Nair et al. 2006) potentiated conditioned reinstatement of cocaine seeking (Feltenstein and See 2006).

1.3.2 INTERACTIVE EFFECTS OF HISTORY OF DEPENDENCE, DRUG CUES, AND STRESS

The above findings document the existence of interactive effects between two factors implicated in craving and relapse: stress and drug-related cues. In view of the significance of a history of dependence in relapse risk, the question arises as to whether an impact of such a drug history on other "triggers" for drug seeking can be documented and studied in animal models. To examine this possibility, a group of animals in the study described above was made dependent via an ethanol vapor procedure following acquisition of operant ethanol self-administration. During the last 2 days of dependence induction, the rats were removed from the vapor chambers each day for 12 h, but allowed to self-administer ethanol paired with response-contingent presentation of an ethanol CS. Rats then were withdrawn from ethanol and subjected to reinstatement tests 3 weeks after termination of the ethanol vapor exposure. In these previously dependent rats, response-contingent presentation of the ethanol CS after footshock substantially enhanced the interactive effects of these stimuli compared to nondependent rats and, in fact, produced synergistic effects on ethanol-seeking (Figures 1.7A,B).

The significance of a dependence history with respect to its role in the effects of alcohol cues and stress is illustrated further by the finding that previously ethanol-dependent rats show enhanced reinstatement induced not only by footshock stress but also by a CS conditioned to footshock stress. But importantly, again, both significant individual and interactive effects of conditioned stress and ethanol cues were observed only in rats with a history of ethanol dependence but not in nondependent rats (Figure 1.5A,B).

1.4 CONCLUDING REMARKS

Persistent vulnerability to relapse represents a formidable challenge for the successful treatment of drug addiction. Understanding of the neurobiological basis of relapse and identification of treatment targets for relapse prevention are issues of particularly high priority in addiction research. Valid and increasingly sophisticated animal models of relapse have become available over the past decade and have been instrumental in expanding our understanding of the neurocircuitry and neural signaling mechanisms that mediate conditioned drug-seeking behavior and contribute to the long-lasting nature of susceptibility to relapse. An important consideration for behavioral strategies in addiction research will be to continue to develop methods that effectively model the diverse aspects of the addiction cycle. The data presented in this chapter document that a history of drug dependence and the experience of negative reinforcement, in addition to the positive-reinforcing, hedonic

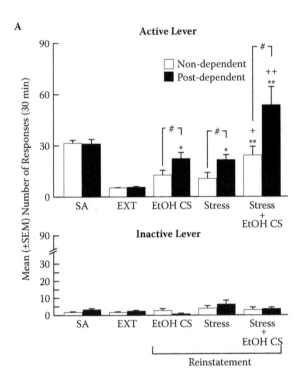

FIGURE 1.7 (A) Responses at a previously active and inactive lever in nondependent and postdependent rats. SA: Ethanol-reinforced responses at the end of self-administration training. EXT: Responses during the final three extinction sessions. Reinstatement: Responses after intermittent footshock (Stress), during response-contingent presentation of an ethanol-associated conditioned stimulus (EtOH CS), and response-contingent presentation of the EtOH CS following footshock (Stress + EtOH CS). $*p < 0.05$, $**p < 0.01$, different from extinction. $+p < 0.05$, $++p < 0.01$, different from Stress and EtOH CS. $\#p < 0.05$ differences between nondependent and postdependent rats.

aspects of the drug experience, convey an additional dimension to the motivating actions of substances of abuse that requires understanding and consideration in medication development. Similarly, the finding that synergistic interactions exist between the response-reinstating effects of drug-associated cues and stress contributes to a better understanding of the relapse process per se by suggesting that the probability of relapse is likely to vary as a function of the number and intensity of risk factors operative at a given time. As well, this finding illustrates that effective pharmacotherapeutic prevention of relapse may require agents that provide concurrent protective effects against multiple risk factors. Using these models, valuable information about agents with such a pharmacological profile are being generated (e.g., Aujla et al. 2007; Baptista et al. 2004; Zhao et al. 2006), information that ultimately will contribute to the development of effective treatment drugs for relapse prevention.

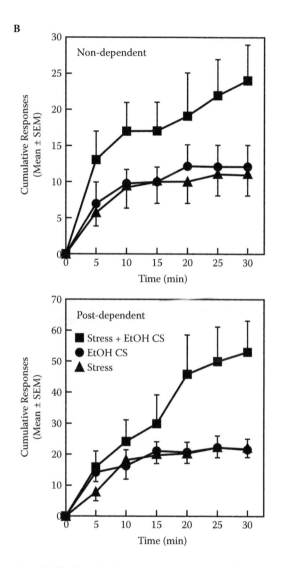

FIGURE 1.7 (continued) (B) Cumulative responses in the Stress, EtOH CS, and Stress + EtOH CS conditions. [Note different ordinate scales for nondependent (top) and postdependent (bottom).] (Reprinted from Liu X, Weiss F (2002a) Additive effect of stress and drug cues on reinstatement of ethanol seeking: exacerbation by history of dependence and role of concurrent activation of corticotropin-releasing factor and opioid mechanisms. *J Neurosci* 22: 7856–7861, with permission.)

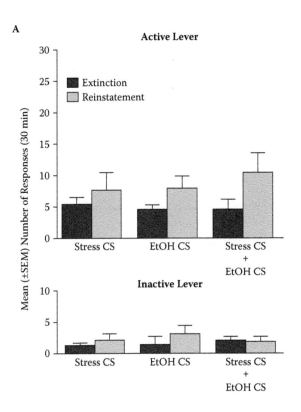

FIGURE 1.8 (A) Responses in ethanol-nondependent rats after exposure to conditioned stress (Stress CS) during response-contingent presentation of an ethanol-associated conditioned stimulus (EtOH CS) and during response-contingent availability of the EtOH CS preceded by exposure to the Stress CS (Stress CS + EtOH CS). Extinction data are presented as the mean (±SEM) across the final three extinction sessions.

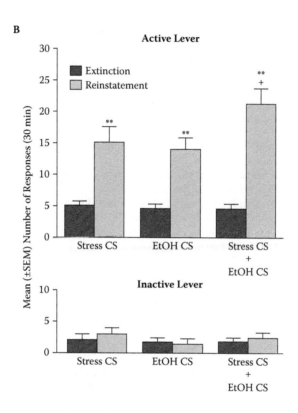

FIGURE 1.8 (continued). (B) Responses at a previously active and an inactive lever in rats with a history of ethanol dependence. Extinction responses were averaged across the final three sessions of the extinction phase. Stress CS: Responses after exposure to conditioned stress; EtOH CS: Responses during response-contingent presentation of an ethanol-associated conditioned stimulus; Stress CS + EtOH CS: Lever presses during response-contingent availability of the EtOH CS following exposure to the Stress CS. **$p < 0.01$, different from extinction; +$p < 0.05$, different from Stress CS and EtOH CS. (From Liu X, Weiss F (2003) Stimulus conditioned to foot-shock stress reinstates alcohol-seeking behavior in an animal model of relapse. *Psychopharmacology* (Berl) 168: 184–191, with permission.)

REFERENCES

Angres DH, Benson WH (1985) Cocainism—a workable model for recovery. *Psychiatr Med* 3: 369–388.

Aronson TA, Craig TJ (1986) Cocaine precipitation of panic disorder. *Am J Psychiatry* 143: 643–645.

Aujla H, Martin-Fardon R, Weiss F (2007) Rats with extended access to cocaine exhibit increased stress reactivity and sensitivity to the anxiolytic-like effects of the mGluR 2/3 agonist LY379268 during abstinence. *Neuropsychopharmacology* 33: 1818–1826.

Bachteler D, Economidou D, Danysz W, Ciccocioppo R, Spanagel R (2005) The effects of acamprosate and neramexane on cue-induced reinstatement of ethanol-seeking behavior in rat. *Neuropsychopharmacology* 30: 1104–1110.

Baptista MA, Martin-Fardon R, Weiss F (2004) Preferential effects of the metabotropic glutamate 2/3 receptor agonist LY379268 on conditioned reinstatement versus primary reinforcement: comparison between cocaine and a potent conventional reinforcer. *J Neurosci* 24: 4723–4727.

Bienkowski P, Kostowski W, Koros E (1999) Ethanol-reinforced behaviour in the rat: effects of naltrexone. *Eur J Pharmacol* 374: 321–327.

Blanchard DC, Blanchard RJ (1999) Cocaine potentiates defensive behaviors related to fear and anxiety. *Neurosci Biobehav Rev* 23: 981–991.

Blanchard RJ, Hori K, Tom P, Blanchard C (1987) Social structure and ethanol consumption in laboratory rats. *Pharmacol Biochem Behav* 28: 437–442.

Bouton ME, Bolles RC (1979) Role of conditioned contextual stimuli in reinstatement of extinguished fear. *J Exp Psychol Anim Behav Process* 5: 368–378.

Bouton ME, Swartzentruber D (1986) Analysis of the associative and occasion-setting properties of contexts participating in a Pavlovian discrimination. *J Exp Psychol Anim Behav Process* 12: 333–350.

Braus DF, Wrase J, Grusser S, Hermann D, Ruf M, Flor H, Mann K, Heinz A (2001) Alcohol-associated stimuli activate the ventral striatum in abstinent alcoholics. *J Neural Transm* 108: 887–894.

Breese GR, Knapp DJ, Overstreet DH (2004) Stress sensitization of ethanol withdrawal-induced reduction in social interaction: inhibition by CRF-1 and benzodiazepine receptor antagonists and a 5-HT1A-receptor agonist. *Neuropsychopharmacology* 29: 470–482.

Brown SA, Vik PW, Patterson TL, Grant I, Schuckit MA (1995) Stress, vulnerability and adult alcohol relapse. *J Stud Alcohol* 56: 538–545.

Burattini C, Gill TM, Aicardi G, Janak PH (2006) The ethanol self-administration context as a reinstatement cue: acute effects of naltrexone. *Neuroscience* 139: 877–87.

Childress AR, Hole AV, Ehrman RN, Robbins SJ, McLellan AT, O'Brien CP (1993) Cue reactivity and cue reactivity interventions in drug dependence. *NIDA Res Monogr* 137: 73–95.

Ciccocioppo R, Angeletti S, Weiss F (2001a) Long-lasting resistance to extinction of response reinstatement induced by ethanol-related stimuli: role of genetic ethanol preference. *Alcohol Clin Exp Res* 25: 1414–1419.

Ciccocioppo R, Economidou D, Fedeli A, Angeletti S, Weiss F, Heilig M, Massi M (2004a) Attenuation of ethanol self-administration and of conditioned reinstatement of alcohol-seeking behaviour by the antiopioid peptide nociceptin/orphanin FQ in alcohol-preferring rats. *Psychopharmacology* (Berl) 172: 170–178.

Ciccocioppo R, Lin D, Martin-Fardon R, Weiss F (2003) Reinstatement of ethanol-seeking behavior by drug cues following single versus multiple ethanol intoxication in the rat: effects of naltrexone. *Psychopharmacology* (Berl) 168: 208–215.

Ciccocioppo R, Martin-Fardon R, Weiss F (2002) Effect of selective blockade of mu(1) or delta opioid receptors on reinstatement of alcohol-seeking behavior by drug-associated stimuli in rats. *Neuropsychopharmacology* 27: 391–399.

Ciccocioppo R, Martin-Fardon R, Weiss F (2004b) Stimuli associated with a single cocaine experience elicit long-lasting cocaine-seeking. *Nat Neurosci* 7: 495–496.

Ciccocioppo R, Sanna PP, Weiss F (2001b) Cocaine-predictive stimulus induces drug-seeking behavior and neural activation in limbic brain regions after multiple months of abstinence: reversal by D(1) antagonists. *Proc Natl Acad Sci U S A* 98: 1976–1981.

Cooney NL, Gillespie RA, Baker LH, Kaplan RF (1987) Cognitive changes after alcohol cue exposure. *J Consult Clin Psychol* 55: 150–155.

Cooney NL, Litt MD, Morse PA, Bauer LO, Gaupp L (1997) Alcohol cue reactivity, negative-mood reactivity, and relapse in treated alcoholic men. *J Abnorm Psychol* 106: 243–250.

Dayas CV, Liu X, Simms JA, Weiss F (2007) Distinct patterns of neural activation associated with ethanol seeking: effects of naltrexone. *Biol Psychiatry* 61: 979–989.

Drummond DC (2001) Theories of drug craving, ancient and modern. *Addiction* 96: 33–46.

Eriksen L, Gotestam KG (1984) Conditioned abstinence in alcoholics: A controlled experiment. *Int J Addict* 19: 287–294.

Everitt BJ, Dickinson A, Robbins TW (2001) The neuropsychological basis of addictive behaviour. *Brain Res Brain Res Rev* 36: 129–138.

Feltenstein MW, See RE (2006) Potentiation of cue-induced reinstatement of cocaine-seeking in rats by the anxiogenic drug yohimbine. *Behav Brain Res* 174: 1–8.

Gawin FH, Kleber HD (1986) Abstinence symptomatology and psychiatric diagnosis in cocaine abusers. *Arch Gen Psychiatry* 43: 107–113.

George MS, Anton RF, Bloomer C, Teneback C, Drobes DJ, Lorberbaum JP, Nahas Z, Vincent DJ (2001) Activation of prefrontal cortex and anterior thalamus in alcoholic subjects on exposure to alcohol-specific cues. *Arch Gen Psychiatry* 58: 345–352.

Ghitza UE, Gray SM, Epstein DH, Rice KC, Shaham Y (2006) The anxiogenic drug yohimbine reinstates palatable food seeking in a rat relapse model: a role of CRF1 receptors. *Neuropsychopharmacology* 31: 2188–2196.

Goeders NE, Guerin GF (1994) Non-contingent electric shock facilitates the acquisition of intravenous cocaine self-administration in rats. *Psychopharmacology* 114: 63–70.

Goodwin RD, Stayner DA, Chinman MJ, Wu P, Tebes JK, Davidson L (2002) The relationship between anxiety and substance use disorders among individuals with severe affective disorders. *Compr Psychiatry* 43: 245–252.

Gracy KN, Dankiewicz LA, Weiss F, Koob GF (2000) Heroin-specific cues reinstate heroin-seeking behavior in rats after prolonged extinction. *Pharmacol Biochem Behav* 65: 489–494.

Greeley JD, Swift W, Prescott J, Heather N (1993) Reactivity to alcohol-related cues in heavy and light drinkers. *J Stud Alcohol* 54: 359–368.

Grimm JW, Hope BT, Wise RA, Shaham Y (2001) Neuroadaptation. Incubation of cocaine craving after withdrawal. *Nature* 412: 141–142.

Haney M, Maccari S, Le Moal M, Simon H, Piazza PV (1995) Social stress increases the acquisition of cocaine self-administration in male and female rats. *Brain Res* 698: 46–52.

Hershon HI (1977) Alcohol withdrawal symptoms and drinking behavior. *J Stud Alcohol* 38: 953–971.

Higley JD, Hasert MF, Suomi SJ, Linnoila M (1991) Nonhuman primate model of alcohol abuse: effect of early experience, personality, and stress on alcohol consumption. *Proc Natl Acad Sci USA* 88: 7261–7265.

Ingjaldsson JT, Thayer JF, Laberg JC (2003) Craving for alcohol and pre-attentive processing of alcohol stimuli. *Int J Psychophysiol* 49: 29–39.

Kajdasz DK, Moore JW, Donepudi H, Cochrane CE, Malcolm RJ (1999) Cardiac and mood-related changes during short-term abstinence from crack cocaine: the identification of possible withdrawal phenomena. *Am J Drug Alcohol Abuse* 25: 629–637.

Kampman KM, Volpicelli JR, Mulvaney F, Alterman AI, Cornish J, Gariti P, Cnaan A, et al. (2001) Effectiveness of propranolol for cocaine dependence treatment may depend on cocaine withdrawal symptom severity. *Drug Alcohol Depend* 63: 69–78.

Kaplan RF, Cooney NL, Baker LH, Gillespie RA, Meyer RE, Pomerleau OF (1985) Reactivity to alcohol-related cues: Physiological and subjective responses in alcoholics and non-problem drinkers. *J Stud Alcohol* 46: 267–272.

Kareken DA, Claus ED, Sabri M, Dzemidzic M, Kosobud AE, Radnovich AJ, Hector D, et al. (2004) Alcohol-related olfactory cues activate the nucleus accumbens and ventral tegmental area in high-risk drinkers: preliminary findings. *Alcohol Clin Exp Res* 28: 550–557.

Kasarabada ND, Anglin MD, Khalsa-Denison E, Paredes A (1998) Variations in psychosocial functioning associated with patterns of progression in cocaine-dependent men. *Addict Behav* 23: 179–189.

Katner SN, Magalong JG, Weiss F (1999) Reinstatement of alcohol-seeking behavior by drug-associated discriminative stimuli after prolonged extinction in the rat. *Neuropsychopharmacology* 20: 471–479.

Katner SN, Weiss F (1999) Ethanol-associated olfactory stimuli reinstate ethanol-seeking behavior after extinction and modify extracellular dopamine levels in the nucleus accumbens. *Alcohol Clin Exp Res* 23: 1751–1760.

Katz JL, Higgins ST (2003) The validity of the reinstatement model of craving and relapse to drug use. *Psychopharmacology* (Berl) 168: 21–30.

Knapp DJ, Overstreet DH, Moy SS, Breese GR (2004) SB242084, flumazenil, and CRA1000 block ethanol withdrawal-induced anxiety in rats. *Alcohol* 32: 101–111.

Kowatch RA, Schnoll SS, Knisely JS, Green D, Elswick RK (1992) Electroencephalographic sleep and mood during cocaine withdrawal. *J Addict Diseases* 11: 21–45.

Kreek MJ (1987) Multiple drug abuse patterns and medical consequences. In: Meltzer HY (ed) *Psychopharmacology: The Third Generation of Progress*. Raven Press, New York, pp 1597–1604.

Kreek MJ, Koob GF (1998) Drug dependence: stress and dysregulation of brain reward pathways. *Drug Alcohol Depend* 51: 23–47.

Laberg JC (1986) Alcohol and expectancy: Subjective, psychophysiological and behavioral responses to alcohol stimuli in severely, moderately and non-dependent drinkers. *Br J Addict* 81: 797–808.

Le A, Shaham Y (2002) Neurobiology of relapse to alcohol in rats. *Pharmacol Ther* 94: 137–156.

Leshner AI (1997) Addiction is a brain disease, and it matters. *Science* 278: 45–47.

Littleton J (2000) Can craving be modeled in animals? The relapse prevention perspective. *Addiction* 95 Suppl 2: S83–S90.

Liu X, Weiss F (2002a) Additive effect of stress and drug cues on reinstatement of ethanol seeking: exacerbation by history of dependence and role of concurrent activation of corticotropin-releasing factor and opioid mechanisms. *J Neurosci* 22: 7856–7861.

Liu X, Weiss F (2002b) Reversal of ethanol-seeking behavior by D1 and D2 antagonists in an animal model of relapse: differences in antagonist potency in previously ethanol-dependent versus nondependent rats. *J Pharmacol Exp Ther* 300: 882–889.

Liu X, Weiss F (2003) Stimulus conditioned to foot-shock stress reinstates alcohol-seeking behavior in an animal model of relapse. *Psychopharmacology* (Berl) 168: 184–191.

Lu L, Grimm JW, Dempsey J, Shaham Y (2004) Cocaine seeking over extended withdrawal periods in rats: different time courses of responding induced by cocaine cues versus cocaine priming over the first 6 months. *Psychopharmacology* (Berl) 176: 101–108.

Macey DJ, Schulteis G, Heinrichs SC, Koob GF (1996) Time-dependent quantifiable withdrawal from ethanol in the rat: Effect of method of dependence induction. *Alcohol* 13: 163–170.

Marinelli PW, Funk D, Juzytsch W, Harding S, Rice KC, Shaham Y, Le AD (2007) The CRF1 receptor antagonist antalarmin attenuates yohimbine-induced increases in operant alcohol self-administration and reinstatement of alcohol seeking in rats. *Psychopharmacology* (Berl) 195: 345–355.

Marlatt GA (1985) Relapse prevention: introduction and overview of the model. In: Marlatt GA, Gordon JR (eds) *Relapse Prevention: Maintenance Strategies in the Treatment of Addictive Behaviors*. Guilford, London, pp 3–70.

Martin WA, Jasinski DR (1969) Physiological parameters of morphine dependence in man—tolerance, early abstinence, protracted abstinence. *J Psychiatr Res* 7: 7–9.

McDougle CJ, Black JE, Malison RT, Zimmermann RC, Kosten TR, Heninger GR, Price LH (1994) Noradrenergic dysregulation during discontinuation of cocaine use in addicts. *Arch Gen Psychiatry* 51: 713–719.

McKay JR, Rutherford MJ, Alterman AI, Cacciola JS, Kaplan MR (1995) An examination of the cocaine relapse process. *Drug Alcohol Dep* 38: 35–43.

McLellan AT, Lewis DC, O'Brien CP, Kleber HD (2000) Drug dependence, a chronic medical illness: implications for treatment, insurance, and outcomes evaluation. *JAMA* 284: 1689–1695.

Meyer RE (1996) The disease called addiction: emerging evidence in a 200 year debate. *Lancet* 347: 162–166.

Miller NS, Gold MS (1994) Dissociation of "conscious desire" (craving) from and relapse in alcohol and cocaine dependence. *Ann Clin Psychiatry* 6: 99–106.

Mollenauer S, Bryson R, Robinson M, Sardo J, Coleman C (1993) EtOH self-administration in anticipation of noise stress in C57BL/6J mice. *Pharmacol Biochem Behav* 46: 35–38.

Monti PM, Binkoff JA, Abrams DB, Zwick WR, Nirenberg TD, Liepman MR (1987) Reactivity of alcoholics and nonalcoholics to drinking cues. *J Abnorm Psychol* 96: 122–126.

Monti PM, Rohsenow DJ, Hutchison KE, Swift RM, Mueller TI, Colby SM, Brown RA, Gulliver SB, Gordon A, Abrams DB (1999) Naltrexone's effect on cue-elicited craving among alcoholics in treatment. *Alcohol Clin Exp Res* 23: 1386–1394.

Monti PM, Rohsenow DJ, Rubonis AV, Niaura RS, Sirota AD, Colby SM, Abrams DB (1993) Alcohol cue reactivity: effects of detoxification and extended exposure. *J Stud Alcohol* 54: 235–245.

Myrick H, Anton RF, Li X, Henderson S, Drobes D, Voronin K, George MS (2004) Differential brain activity in alcoholics and social drinkers to alcohol cues: relationship to craving. *Neuropsychopharmacology* 29: 393–402.

Nair SG, Gray SM, Ghitza UE (2006) Role of food type in yohimbine- and pellet-priming-induced reinstatement of food seeking. *Physiol Behav* 88: 559–566.

Nash J, Jr., Maickel RP (1988) The role of the hypothalamic-pituitary-adrenocortical axis in post-stress induced ethanol consumption by rats. *Prog Neuropsychopharmacol Biol Psychiatry* 12: 653–671.

O'Brien CP, Childress AR, Ehrman R, Robbins SJ (1998) Conditioning factors in drug abuse: Can they explain compulsion? *J Psychopharmacol* 12: 15–22.

O'Brien CP, McLellan AT (1996) Myths about the treatment of addiction. *Lancet* 347: 237–240.

O'Brien CP, Volpicelli LA, Volpicelli JR (1996) Naltrexone in the treatment of alcoholism: a clinical review. *Alcohol* 13: 35–39.

Overstreet DH, Knapp DJ, Breese GR (2004) Modulation of multiple ethanol withdrawal-induced anxiety-like behavior by CRF and CRF1 receptors. *Pharmacol Biochem Behav* 77: 405–413.

Ramsey NF, Van Ree M (1993) Emotional but not physical stress enhances intravenous cocaine self-administration in drug naive rats. *Brain Res* 608: 216–222.

Rassnick S, Heinrichs SC, Britton KT, Koob GF (1993) Microinjection of a corticotropin-releasing factor antagonist into the central nucleus of the amygdala reverses anxiogenic-like effects of ethanol withdrawal. *Brain Res* 605: 25–32.

Razzouk D, Bordin IA, Jorge MR (2000) Comorbidity and global functioning (DSM-III-R Axis V) in a Brazilian sample of cocaine users. *Subst Use Misuse* 35: 1307–1315.

Roberts AJ, Cole M, Koob GF (1996) Intra-amygdala muscimol decreases operant ethanol self-administration in dependent rats. *Alcohol Clin Exp Res* 20: 1289–1298.

Rohsenow DJ, Monti PM, Hutchison KE, Swift RM, Colby SM, Kaplan GB (2000) Naltrexone's effects on reactivity to alcohol cues among alcoholic men. *J Abnorm Psychol* 109: 738–742.

Rounsaville BJ, Anton SF, Carroll K, Budde D, Prusoff BA, Gawin F (1991) Psychiatric diagnoses of treatment-seeking cocaine abusers. *Arch Gen Psychiatry* 48: 43–51.

Sarnyai Z, Shaham Y, Heinrichs SC (2001) The role of corticotropin-releasing factor in drug addiction. *Pharmacol Rev* 53: 209–243.

Satel SL, Kosten TR, Schuckit MA, Fischman MW (1993) Should protracted withdrawal from drugs be included in DSM-IV? *Am J Psychiatry* 150: 695–704.

Schneider F, Habel U, Wagner M, Franke P, Salloum JB, Shah NJ, Toni I, Sulzbach C, Honig K, Maier W, Gaebel W, Zilles K (2001) Subcortical correlates of craving in recently abstinent alcoholic patients. *Am J Psychiatry* 158: 1075–1083.

Schulteis G, Markou A, Cole M, Koob GF (1995) Decreased brain reward produced by ethanol withdrawal. *Proc Natl Acad Sci USA* 92: 5880–5884.

See RE (2002) Neural substrates of conditioned-cued relapse to drug-seeking behavior. *Pharmacol Biochem Behav* 71: 517–529.

See RE (2005) Neural substrates of cocaine-cue associations that trigger relapse. *Eur J Pharmacol* 526: 140–146.

See RE, Grimm JW, Kruzich PJ, Rustay N (1999) The importance of a compound stimulus in conditioned drug-seeking behavior following one week of extinction from self-administered cocaine in rats. *Drug Alcohol Depend* 57: 41–49.

Shaham Y, Erb S, Stewart J (2000) Stress-induced relapse to heroin and cocaine seeking in rats: a review. *Brain Res Brain Res Rev* 33: 13–33.

Shaham Y, Miczek KA (2003) Reinstatement—toward a model of relapse. *Psychopharmacology* (Berl) 168: 1–2.

Shaham Y, Shalev U, Lu L, De Wit H, Stewart J (2003) The reinstatement model of drug relapse: history, methodology and major findings. *Psychopharmacology* (Berl) 168: 3–20.

Shalev U, Grimm JW, Shaham Y (2002) Neurobiology of relapse to heroin and cocaine seeking: a review. *Pharmacol Rev* 54: 1–42.

Sinha R (2000) Stress, craving, and relapse to drug use. National Institute on Drug Abuse (NIDA) Symposium: The Intersection of Stress, Drug Abuse, and Development, Bethesda, MD.

Sinha R (2001) How does stress increase risk of drug abuse and relapse? *Psychopharmacology* (Berl) 158: 343–359.

Sinha R, Catapano D, O'Malley S (1999) Stress-induced craving and stress response in cocaine dependent individuals. *Psychopharmacology* (Berl) 142: 343–351.

Sinha R, Fuse T, Aubin LR, O'Malley SS (2000) Psychological stress, drug-related cues and cocaine craving. *Psychopharmacology* (Berl) 152: 140–148.

Sinha R, Garcia M, Paliwal P, Kreek MJ, Rounsaville BJ (2006) Stress-induced cocaine craving and hypothalamic-pituitary-adrenal responses are predictive of cocaine relapse outcomes. *Arch Gen Psychiatry* 63: 324–331.

Sinha R, Talih M, Malison R, Cooney N, Anderson GM, Kreek MJ (2003) Hypothalamic-pituitary-adrenal axis and sympatho-adreno-medullary responses during stress-induced and drug cue-induced cocaine craving states. *Psychopharmacology* (Berl) 170: 62–72.

Stormark KM, Laberg JC, Bjerland T, Nordby H, Hugdahl K (1995) Autonomic cued reactivity in alcoholics: the effect of olfactory stimuli. *Addict Behav* 20: 571–584.

Streeter CC, Gulliver SB, Baker E, Blank SR, Meyer AA, Ciraulo DA, Renshaw PF (2002) Videotaped cue for urge to drink alcohol. *Alcohol Clin Exp Res* 26: 627–634.

Tiffany ST, Carter BL (1998) Is craving the source of compulsive drug use? *J Psychopharmacol* 12: 23–30.

Tiffany ST, Carter BL, Singleton EG (2000) Challenges in the manipulation, assessment and interpretation of craving relevant variables. *Addiction* 95 Suppl 2: S177–187.

Tiffany ST, Conklin CA (2000) A cognitive processing model of alcohol craving and compulsive alcohol use. *Addiction* 95 Suppl 2: S145–153.

Valdez GR, Roberts AJ, Chan K, Davis H, Brennan M, Zorrilla EP, Koob GF (2002) Increased ethanol self-administration and anxiety-like behavior during acute ethanol withdrawal and protracted abstinence: regulation by corticotropin-releasing factor. *Alcohol Clin Exp Res* 26: 1494–1501.

Valdez GR, Zorrilla EP, Rivier J, Vale WW, Koob GF (2003) Locomotor suppressive and anxiolytic-like effects of urocortin 3, a highly selective type 2 corticotropin-releasing factor agonist. *Brain Res* 980: 206–212.

Van De Laar MC, Licht R, Franken IH, Hendriks VM (2004) Event-related potentials indicate motivational relevance of cocaine cues in abstinent cocaine addicts. *Psychopharmacology* (Berl) 177:121–129.

Volpicelli JR, Alterman AI, Hayashida M, O'Brien CP (1992) Naltrexone in the treatment of alcohol dependence. *Arch Gen Psychiatry* 49: 876–880.

Walfish S, Massey R, Krone A (1990) Anxiety and anger among abusers of different substances. *Drug Alcohol Depend* 25: 253–256.

Wallace BC (1989) Psychological and environmental determinants of relapse in crack cocaine smokers. *J Subst Abuse Treat* 6: 95–106.

Weiss F (2005) Neurobiology of craving, conditioned reward and relapse. *Curr Opin Pharmacol* 5: 9–19.

Weiss F, Maldonado-Vlaar CS, Parsons LH, Kerr TM, Smith DL, Ben-Shahar O (2000) Control of cocaine-seeking behavior by drug-associated stimuli in rats: effects on recovery of extinguished operant-responding and extracellular dopamine levels in amygdala and nucleus accumbens. *Proc Natl Acad Sci U S A* 97: 4321–4326.

Weiss F, Martin-Fardon R, Ciccocioppo R, Kerr TM, Smith DL, Ben-Shahar O (2001) Enduring resistance to extinction of cocaine-seeking behavior induced by drug-related cues. *Neuropsychopharmacology* 25: 361–372.

Zhao Y, Dayas CV, Aujla H, Baptista MA, Martin-Fardon R, Weiss F (2006) Activation of group II metabotropic glutamate receptors attenuates both stress and cue-induced ethanol-seeking and modulates c-fos expression in the hippocampus and amygdala. *J Neurosci* 26: 9967–9974.

Ziedonis DM, Rayford BS, Bryant KJ, Rounsaville BJ (1994) Psychiatric comorbidity in white and African-American cocaine addicts seeking substance abuse treatment. *Hosp Community Psychiatry* 45: 43–49.

2 Application of Chronic Extracellular Recording Method to Studies of Cocaine Self-Administration
Method and Progress

*Laura L. Peoples, Alexai V. Kravitz,
and Karine Guillem*

CONTENTS

2.1 CHAPTER OVERVIEW

To further the investigation of the neurobiology of drug addiction, researchers have integrated chronic extracellular recording and intravenous drug self-administration procedures. The technique allows researchers to characterize the discharge patterns of single neurons during drug-directed instrumental behavior. This method is the topic of the present chapter. The first sections of the chapter overview the method and discuss the rationale, advantages, and disadvantages of the technique. These sections also describe some of the experimental approaches that are required for certain applications of the recording procedure (see Sections 2.2–2.4). The remainder of the chapter will review three areas of research, as follows: (1) firing patterns of NAc neurons during drug-directed behavior (Sections 2.5–2.6); (2) acute effects of cocaine on NAc activity (Section 2.7); and (3) effects of repeated cocaine self-administration on NAc firing (Section 2.8). The research exemplifies some of the experimental methods discussed in earlier parts of the chapter, as well as some of the novel research opportunities that are afforded by the technique. The chapter also includes a discussion of future directions. A detailed description of technical aspects of the method is provided in an appendix.

2.2 BASIC METHOD

The intravenous drug self-administration model is a well-validated model of human drug self-administration (Johanson and Balster 1978; Griffiths et al. 1980; Stewart et al. 1984; O'Brien et al. 1998). Under appropriate conditions of drug exposure, the method is also expected to serve as a valid model of drug addiction (Shalev et al. 2002; Shaham et al. 2003; Panlilio and Goldberg 2007). In the simplest case, animals are implanted with an intravenous catheter and then trained to engage in operant behavior, typically a lever press response, according to a Fixed-Ratio 1 (FR1) schedule of reinforcement. Each press of the lever is followed by a single infusion of drug, which is typically paired with a conditioned reinforcer such as a light or a tone. Although the experimenter controls the drug concentration and volume of each infusion, animals control the timing of infusions. Within the confines of drug availability, humans also control their own drug exposure. Interestingly the patterns of intake exhibited by animals in the self-administration paradigm are quite similar to those of humans. Drug exposure is an important determinant of both acute and chronic drug effects. Use of the self-administration procedure increases the likelihood that animals are exposed to levels and patterns of drug exposure that are consistent with those of drug addicts. Researchers using the self-administration paradigm are thus more likely to characterize drug effects that are relevant to drug addiction, as compared to experimenter-determined injections.

If exposed to short daily sessions and moderate drug doses, animals exhibit stable rates of intake across many days. If, on the other hand, animals are exposed to high doses of drug and/or long daily sessions, animals exhibit escalation of drug intake across repeated self-administration sessions, as well as other behaviors that are considered addiction-like (Koob et al. 1989; Ahmed and Koob 1998;

Koob and Le Moal 2001, 2008; Deroche-Gamonet et al. 2004; Vanderschuren and Everitt 2004). The emergence of these addiction-like behaviors provides additional evidence of the validity of the self-administration paradigm as a model for studies of drug addiction. Moreover, brain areas implicated in mediating drug self-administration in animals are consistent with regions implicated in drug reward and addiction in humans (Grant et al. 1996; Breiter et al. 1997; Childress et al. 1999; Volkow et al. 1999). Identification of the neural mechanisms that mediate drug self-administration behavior is expected to contribute to delineating causes of drug addiction.

In chronic extracellular recording studies of drug self-administration (referred to herein as chronic behaving animal recording studies), animals are chronically implanted with either a bundle or an array of insulated microwires (e.g., 50 μm dia stainless steel wires with Teflon-coating) (Figure 2.1). During recording sessions, a headstage containing electrical contacts to the microwires is connected via a flexible cable to a recording system. Animals can move freely and exhibit typical operant behavior. During recording sessions, voltage signals from each microwire pass through an amplification and filtering system (Figures 2.2–2.3). Signals that meet preselected criteria (e.g., amplitude criteria) are time stamped and stored for subsequent offline analysis. After the experiment, analysis software is used to discriminate voltage signals that correspond to action potentials of individual neurons (Figure 2.4). Histogram analyses are then used to test for changes in firing rate of single neurons in relation to particular events (Figure 2.5). A more detailed description of the technique is included in an appendix at the end of this chapter. Additional details, as well as a number of equipment photographs, are included in (Peoples 2003).

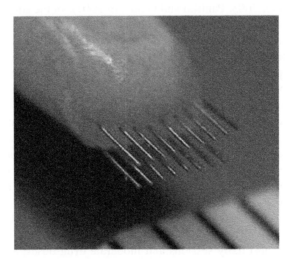

FIGURE 2.1   Microwire array. Photograph of the tip of a 16-wire microwire array. The array is 2mm in width and 1mm in height. Photograph shows wire tips emerging from polyethylene glycol, which is necessary for maintaining the configuration of the wires but is slowly dissolved as wires are lowered into brain (see Appendix for details). Array was purchased from MicroProbe, Inc.

FIGURE 2.2 Rat engaged in intravenous cocaine self-administration during a chronic extracellular recording session. Photograph shows a rat pressing the operant lever during a cocaine self-administration session. The headstage of the animal is connected to an electronic cable that feeds to a fluid-electronic swivel. The intravenous catheter is connected to the same swivel by spring-covered tubing. Figure with permission from Peoples, Application of chronic extracellular recording to studies of drug self-administration. In: *Methods in Drug abuse research: cellular and circuits level analyses*, pp 161–211: Waterhouse, B.D., Ed., Boca Raton, FL: CRC Press, 2003.

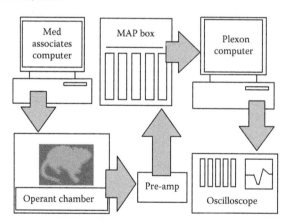

FIGURE 2.3 Basic electrophysiological hardware setup. Med Associates operant hardware and software are integrated with the Plexon electrophysiological recording system. Microwire signals pass through a headset amplifier mounted on the rat's head to a differential recording preamp. The signal on each microwire is then individually filtered and amplified by the Multi-Unit Acquisition Processor (MAP Box, Plexon, Inc.). Neural recordings are inspected on an oscilloscope, and digitally recorded on a dedicated computer during the operant behavioral sessions. Time-stamped digitized video feed along with time-stamp records of behavioral and electrophysiological events are time-synchronized so that we can characterize the firing of single accumbal neurons in relation to any stimulus and behavioral events that are tracked by either the Med Associates hardware and software, or by experimenter analysis of video.

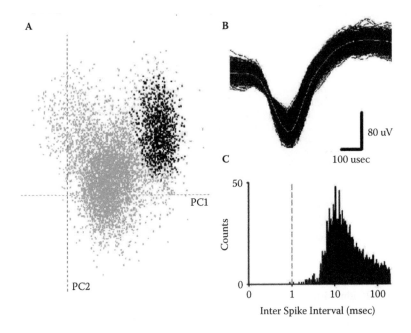

FIGURE 2.4 Example of extracellular recordings of single-neuron discharges. A. Scatterplot of the first two principal components (PC1 and PC2) of all large-amplitude electrical events recorded from one microwire during 15 minutes of an operant session. Events with similar shapes cluster together in this plot and can thereby be categorized as different clusters. Typically, there is a cluster of events that cannot be resolved into single-unit activity and is referred to as "noise" (gray dots), and one or more well-separated clusters that correspond to activity from a single unit (black dots). B. All waveforms from the black cluster in A are superimposed over each other, demonstrating that they all have a similar shape, consistent with an extracellularly recorded neural waveform. White line represents the average of these waveforms. C. Interspike interval graph of the waveforms in B. Very few intervals were less than 1 msec, indicating that the spikes arose from a single unit.

The occurrence of action potentials can be tracked with a 1 msec resolution. The recordings are highly stable such that the activity of a single neuron can be followed for many hours and sometimes for multiple days (Peoples et al. 1999a; Janak 2002). It is thus possible to characterize changes in neuronal firing that occur time-locked to rapid discrete events, such as the cocaine-reinforced operant response (i.e., msec to sec time frame). One can also concurrently evaluate slow and long-lasting changes in firing that might be associated with changes in either drug exposure or the motivational state of the animal (i.e., time frame of min to hrs).

With histological procedures, the location of recording wires can be identified with a resolution of ≈100 µm) (Uzwiak et al. 1997). It is thus possible to use the procedure to test for differences in single neuron activity across relatively small subregions of a given structure (Uzwiak et al. 1997; Hollander and Carelli 2005; Peoples et al. 2007b).

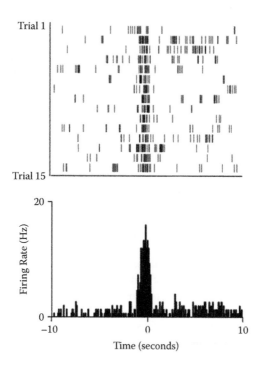

FIGURE 2.5 Example of a raster and histogram display for a single neuron. Top: A raster display marks the occurrence of each discharge of a particular neuron for 15 cocaine-reinforced lever presses. Each tick corresponds to a single discharge. Each row shows all the discharges during the 10 sec before and after each of the 15 lever presses. Bottom: A histogram shows average number of discharges per sec (Hertz, Hz) calculated as a function of 0.1-sec bins. Average firing was calculated for the same 15 trials shown in the raster display.

2.3 RATIONALE, ADVANTAGES, AND DISADVANTAGES OF THE METHOD

2.3.1 IMPORTANCE OF ELECTROPHYSIOLOGICAL INVESTIGATIONS FOR IDENTIFYING CAUSES AND TREATMENTS OF ADDICTION

Drug addiction is a progressive and chronic disorder characterized by the emergence of compulsive drug seeking and taking. It is also associated with a concomitant weakening of other motivated behaviors (Gawin and Kleber 1986; Weddington et al. 1990; American Psychiatric Association 1994). It is hypothesized that the maladaptive behavior that characterizes addiction is mediated by long-lasting drug-induced neuroadaptations within neuronal circuits that control motivated behavior (Berke and Hyman 2000; Everitt and Wolf 2002; Robinson and Berridge 2003; Kalivas 2004; Thomas et al. 2008). Identifying the neuroadaptations and the drug actions that mediate the adaptations are necessary steps for understanding the causes of addiction as well as for the development of effective pharmacotherapies.

Drug-induced adaptations involve changes in protein function in individual neurons, which can impact the excitability and activity of the individual neurons. Changes in single neuron firing can alter the activity of neuronal circuits and lead to changes in behavior. Understanding the specific alterations in protein function is a critical aspect of identifying drug-induced adaptations that might contribute to drug addiction. However, it is also critical to identify the effects of drug on neuronal activity.

Addictive drugs induce a plethora of changes in protein synthesis and function (Nestler 2001; Self et al. 2004; Hope et al. 2005; Lu et al. 2005; Ben-Shahar et al. 2007; Zhang et al. 2007; Kalivas and O'Brien 2008). Understanding the impact of drug exposure on the excitability and activity of individual neurons can help to differentiate among changes in protein function that might be of greatest functional consequence. Additionally, it is possible that mechanisms that mediate addiction involve changes in the activity of neuronal microcircuits, as well as regional circuits. Knowledge of drug-induced changes in single neuron and circuit function, combined with an understanding of the drug-induced changes in protein function that underlie the neurophysiological plasticity, is necessary for a full understanding of mechanisms that mediate the development of addiction. The multilevel characterization of drug-induced plasticity is also expected to identify a range of potential targets for pharmacotherapy. These targets may involve primary sites of drug action, but could also involve secondary sites, that might be targeted to compensate or reverse drug-induced changes in protein function, neural excitability, or circuit function.

2.3.2 CHARACTERIZATION OF ACUTE AND CHRONIC DRUG EFFECTS ON SINGLE NEURON ACTIVITY

Identifying drug-induced neurophysiological plasticity and the mechanisms that mediate the plasticity involves characterization of neurophysiological changes in neuronal function that occur as a consequence of repeated drug exposure. However, neuroplasticity is importantly influenced by acute drug actions. Acute pharmacological actions can influence neuroadaptive processes in three ways. First, acute drug actions stimulate receptors and signal transduction pathways. These actions lead to the onset of adaptations in the affected neurons. Second, acute drug actions impact neuron firing rates during the period of drug exposure. These changes in activity can impact receptor-mediated changes in signal transduction pathways in the affected neuron and thereby influence the occurrence of adaptations (Wolf et al. 2004). Third, acute drug-induced changes in neural activity can alter afferent input into other circuits and potentially influence adaptive processes ongoing in those circuits (e.g., drug could alter inputs to memory circuits). A complete understanding of drug-induced neuroadaptations thus depends on characterization of both acute and chronic drug effects.

2.3.3 STUDIES OF ACUTE AND CHRONIC DRUG EFFECTS: ADVANTAGES AND DISADVANTAGES OF DIFFERENT ELECTROPHYSIOLOGICAL APPROACHES

Many studies of acute and chronic effects of cocaine on neuronal activity have employed either slice recording preparations or anesthetized animals. These electrophysiological methods have a number of advantages. The methods can be used to characterize specific biochemical and molecular mechanisms that mediate drug-induced changes in neural activity (White and Kalivas 1998; Zhang et al. 2002; Hu et al. 2004, 2005; Nasif et al. 2005).

Despite the advantages of the anesthetized animal and slice procedures, the techniques have certain limitations. Drug effects on single neuron activity can depend on multiple factors. These include but are not limited to the state of depolarization, the firing rate of a neuron, and the identity of afferent input to a neuron at the time of drug exposure (O'Donnell et al. 1999; Floresco et al. 2003; O'Donnell 2003; West et al. 2003). There is a potential for marked differences in the state of these variables between behaving animals and either the slice or the anesthetized animal preparations. In both the slice and anesthetized animal preparations, individual neurons can be cut off from normal afferent input. Anesthesia also directly impacts the response of neurons to afferent input (Mereu et al. 1983; Yoshimura et al. 1985; Fink-Jensen et al. 1994; Antkowiak and Helfrich-Forster 1998; Antkowiak, 1999). Moreover, particular behaviors are associated with unique patterns of afferent input to individual neurons, which are necessarily absent in the slice and anesthetized animal preparations. Given these differences and the condition-dependent nature of drug actions, it is possible that certain acute and chronic drug effects that occur in the slice and anesthetized animal preparations are not representative of those that occur in the behaving animal. It is also possible that certain effects of drug that occur in the behaving animal and that contribute to addiction are absent in the slice and anesthetized animal preparations.

The potential for different findings between recordings in the slice and anesthetized animal preparations, and the recordings in behaving animals is emphasized by the following: (1) previous demonstrations of such differences (Deadwyler 1986; West et al. 1997; Moxon 1999), and (2) previous observations of behavior-dependent drug effects on various neural substrates (Smith et al. 1980, 1982; Hemby et al. 1997). Additionally, a number of extracellular recording studies during periods of ongoing cocaine seeking and self-administration show that there are differences in acute and chronic cocaine effects that are observed between the behaving animal recording studies and the slice and anesthetized animal preparations (Peoples and Cavanaugh 2003; Peoples et al. 2004, 2007b) (see Sections 2.7–2.8).

Based on the above observations, recordings in animals could be essential for identifying the acute and chronic drug effects on neuronal signaling that mediate the development of addiction.

Electrophysiological studies in behaving animals also provide an additional unique research opportunity. The chronic-behaving animal recording methods can be used to conduct fine-grained correlational analyses of the activity of individual neurons and ongoing sensory and behavioral events. This measurement capacity provides a unique opportunity for testing functional hypotheses of drug addiction (see Sections 2.4.2 and 2.8.4).

Despite the advantages of the chronic-behaving animal recording technique, use of the method in investigations of acute and chronic drug effects is associated with challenges and limitations. In behaving animals, a change in the activity of a single neuron that occurs in response to acute drug exposure can reflect either of the following: (1) pharmacological actions on the recorded neuron, or primary afferent input to the neuron, and (2) an effect of a nonpharmacological factor on the firing of the recorded neuron. In regard to the latter, drug actions in nonrecorded regions can lead to changes in behavior, sensory processes, and motivational processes. Recorded neurons potentially receive afferent input related to those behaviors and processes (i.e., feedback). If this is the case, changes in activity of a neuron during drug exposure can be entirely due to the changes in behavior, sensory, or motivational processes rather than to an effect of drug on the recorded neurons. Similarly, a change in neural activity that is observed in association with repeated exposure to drug self-administration potentially corresponds to either of the following: (1) drug-induced neuroadaptations or (2) normal plasticity associated with repeated exposure to nonpharmacological aspects of the experimental procedures, such as operant conditioning. It is necessary to differentiate the relative contribution of pharmacological and nonpharmacological variables for the findings of the studies to be useful in identifying and investigating drug effects.

With appropriate controls it is possible to identify pharmacologically mediated changes in neural activity in behaving animals. However, characterization of pharmacological mechanisms that contribute to addiction involves not only identification of drug effects but investigation of the specific drug actions that mediate those effects. There are a number of procedures that are typically combined with anesthetized and slice electrophysiological recording techniques to characterize drug actions. Unfortunately, the methods are not readily incorporated with the chronic recording method.

One of these methods involves the use of dye and tracing techniques can be used to identify the specific type of neuron from which neural recordings are made: moreover, the techniques can be used to identify the specific afferent inputs and cellular proteins that are either involved in or affected by drug actions. Another method for investigating the mechanisms of drug action is to use iontophoretic and microinjection techniques to make local pharmacological manipulations of potential sites of drug action. Application of these techniques is difficult in behaving animals but it is especially complicated by the chronic implantation of microwires. The chronic implant of the fine wires prevents mechanical displacement of the wires in brain and is necessary for obtaining the long-duration, stable recordings in the behaving animal. Animals can be very active during drug self-administration sessions and recordings sometimes have to be extended for hours in order to obtain a sufficient sample of neural responses to task-related events such as operant behavior, which can occur at a slow rate during drug self-administration (e.g., at an intermediate dose of cocaine, animals might press the lever once every 6–8 min). With technical development efforts, it is most likely possible to integrate some of the techniques used in other recording studies to the chronic recordings in behaving animals, or to develop alternative methods of investigating mechanisms of drug action in the behaving animal. At the moment, however, opportunities for that type of research are quite limited. Possible avenues for improving this situation are discussed in Future Directions (Section 2.9).

2.3.4   An Integrated Electrophysiological Approach

Given the relative strengths and weaknesses of the different electrophysiological methods, complementary application of the techniques, and consideration of findings from those studies as a whole, will be important to understanding the drug effects that contribute to drug addiction. Slice and anesthetized animal methods can be used to identify drug effects and to characterize mechanisms of drug action. Recordings in behaving animals can be used to test for changes in neuronal activity that are predicted on the basis of the slice and anesthetized recordings, and to test whether the changes in neural activity are associated with the development of addiction-like patterns of behavior. The advantages that can be gained by such an integrative approach are demonstrated by the research described in the present chapter.

2.4   APPLICATION OF THE METHOD: DIFFERENTIATING PHARMACOLOGICAL AND NONPHARMACOLOGICAL EFFECTS AND TESTING ADDICTION HYPOTHESES

2.4.1   Methods for Differentiating between Pharmacological and Nonpharmacological Effects on Neural Activity

2.4.1.1   Acute Drug Effects

Numerous methods can be used to differentiate between a pharmacological effect and a behavioral feedback effect on single neuron activity. One simple strategy for making the differentiation is to compare the time course of the changes in firing that occur after a pharmacological manipulation to the time course of changes in behavior. A dissociation between the time courses shows that the firing rate changes cannot be attributed to behavioral changes. A related strategy is to compare the dose response curves for drug effects on firing and particular behaviors. A difference between the dose response curves for changes in neural activity and behavior can indicate that the two dependent measures are unrelated. A third strategy is to evaluate the effect of manipulating the behavior on neural activity under drug-free conditions. If manipulation of the behavior does not engender a change in firing that mimics the effect of acute drug exposure, then the change in firing observed during drug exposure cannot be attributed to changes in that behavior. A fourth control for the effects of behavioral feedback is a *behavioral clamp*. The goal of the behavioral clamp is to make comparisons of different drug conditions across periods in which behavior is constant. If the originally observed drug-correlated change in firing is unaltered by the behavioral clamp, the neural change cannot be attributed to the behavior (Ranck et al. 1983). A detailed example of this control is included in Section 2.7.

There are several approaches to behavioral clamping. In some cases certain spontaneous behaviors occur during all drug exposure conditions, albeit with different frequencies. If one were concerned that the different frequency with which the behavior occurs was mediating changes in activity of the recorded neuron, one could limit comparisons of firing rate to periods in which that particular behavior occurs, for example, sample firing rate only during periods in which the animal is engaged in locomotion. Making comparisons of firing rates during these periods holds the impact of that behavior constant across the different pharmacological conditions. Another behavioral clamp strategy is to condition animals to engage in a particular behavior under the different drug exposure conditions and to limit firing-rate comparisons to the periods of the conditioned behavior.

In addition to changing behavior per se (e.g., inducing increases in locomotion or stereotypy), the acute effects of addictive drugs have reinforcing properties. Within a limited range, increments in doses increase the strength of these reinforcing properties, consistent with effects of increasing the magnitude of natural rewards (Pickens and Thompson 1968; Johanson and Schuster 1975; Richardson and Roberts 1996; Arroyo et al. 1998; Olmstead et al. 2000; Mantsch et al. 2001; Peltier et al. 2001). The changes in reinforcer magnitude can in turn impact incentive processes and operant behavior. In addition to the primary reinforcing properties, acute effects of addictive drugs, within a limited dose range, can increase the invigorating and conditioned reinforcing properties of conditioned reward-associated stimuli on instrumental behavior (Robbins 1978; Wyvell and Berridge 2000), and can additionally increase Pavlovian conditioned approach behavior (Wan and Peoples 2008). Thus, if one is recording neurons in a region that is potentially involved in tracking aspects of these motivational processes, it is necessary to consider whether changes in neural activity reflect a neuronal response to the altered motivational process, or a more direct effect of drug on the recorded neuron. The approaches that one can use to differentiate between a change in "motivational feedback" and a more direct pharmacological effect are comparable to those described above in reference to behavioral feedback.

2.4.1.2   Chronic Drug Effects

Drug addiction symptoms develop progressively across a history of repeated drug use. Some addicts never attain a significant period of extended abstinence. Others may cycle between periods of active drug use, extended periods of abstinence, and relapse to active drug use. A full understanding of neuroadaptations underlying addiction may depend on characterizing neural activity across these repeated phases of drug exposure, extended abstinence, and re-exposure. Such studies will involve making comparisons of neural activity at different time points across a history of repeated daily self-administration sessions, and in some cases across phases of active drug use, drug abstinence, and drug re-exposure.

With chronically implanted wires, it is possible to follow the activity of individual neurons for a period of days to perhaps two weeks. However, most neurons cannot be reliably recorded for more than several days (Peoples et al. 1999a; Janak 2003). Moreover, studies that aim to characterize effects of cycles of drug self-administration, abstinence, and re-exposure need to maintain recordings for periods of months. A more practical alternative for characterizing chronic drug effects in behaving animal recording studies is to follow a *between-group* approach. That is, one can record separate groups of neurons at the different time points in the drug self-administration history and then test for an effect of repeated drug exposure by comparing average firing of neurons across the different time points. Though one can look at overall, average firing of the population at each time point, it is useful to take advantage of the spatial resolution of the technique and to subtype recorded neurons into groups that exhibit particular types of firing patterns during the self-administration sessions. For example, at each time point, one can identify neurons that show phasic responses to conditioned drug-associated stimuli and then compare the average firing of those different groups of cue-responsive neurons to test for an effect of repeated drug sessions, and so forth. This type of approach has been used successfully in several studies that will be discussed later in the chapter (see Section 2.8).

As is the case for acute drug effects, a challenge in behaving animal studies of chronic drug effects is to differentiate changes in firing that are of a primary pharmacological origin from those that are not. Changes in neural activity across repeated daily self-administration sessions can reflect cocaine-induced neuroadaptations in the recorded neuron. But, they could also reflect normal neural responses to repeated exposure to certain experimental treatments (i.e., neural correlates of learning). To be useful in identifying cocaine-induced neuroadaptations, it is necessary for behaving animal studies to differentiate between these possibilities. There are a number of approaches that can be employed. First, changes in neural activity that are based on processes such as habituation or associative learning should be situation-dependent, showing rapid changes in response to manipulations of various nonpharmacological aspects of the experimental treatment (e.g., the response–reinforcer condition or the operant). Changes in neural activity that reflect long-lasting drug-induced adaptations are expected to be relatively insensitive to similar manipulations. A second and more convincing method of differentiating drug-induced adaptation from experiential effects on neural activity is to compare the changes in firing observed in animals exposed to similar treatments but trained to self-administer a natural reward instead of a drug reward. Given a carefully controlled study, the selective occurrence of neural changes in the drug-exposed animals supports a pharmacological interpretation of the drug data. Examples of these approaches are described later in the chapter (see Section 2.8).

2.4.2 Testing Drug Addiction Hypotheses

Drug addiction is characterized by disruptions in motivated behavior. Individuals exhibit compulsive and uncontrollable drug-seeking and taking, despite knowledge of adverse consequences. Individuals exhibit a weakened ability to use information about adverse consequences to regulate instrumental behavior and a narrowing of behavioral repertoire (i.e., decrease in alternative motivated behaviors). Consistent with these characteristics, addiction is hypothesized to be caused by drug-induced alterations in neural processes that normally regulate motivated behavior.

A number of researchers have put forth specific hypotheses. For example, the incentive sensitization hypothesis postulates that neuroadaptations alter mechanisms that attribute incentive salience to rewards and conditioned reward-associated stimuli. The drug-induced abnormalities are hypothesized to lead to an abnormally strong response of those mechanisms to drug and drug-associated cues (Robinson and Berridge 1993; Koob and Le Moal 1997; Koob and Nestler 1997). The homeostatic dysregulation hypothesis postulates that neuronal mechanisms that regulate affective state and hedonic responses to primary reward are altered so that the baseline affective state of an individual becomes highly negative. Moreover, the hedonic impact of natural rewards is no longer sufficient to normalize the affective state of individuals, though the direct effects of incrementing doses of drug are sufficient to do so (Koob and Le Moal 2001). The stimulus-response (S-R) learning hypothesis proposes that drug-induced adaptations lead to abnormalities in neuronal mechanisms that regulate habits, which are automatic behavioral responses induced by conditioned stimuli. The differential neuroplasticity hypothesis proposes that neurons that regulate natural reward-directed behavior and neurons that facilitate the drug-directed behavior undergo different, activity-dependent neuroadaptations. The differential neuroadaptations tend to weaken neural signaling related to natural reward and to strengthen signaling related to drug, which contributes to corresponding changes in behavior (Peoples and Cavanaugh 2003; Peoples et al. 2004, 2007a, 2007b). In addition to specifying functional disruptions, most hypotheses put forth specific proposals about the underlying brain regions.

All of the addiction hypotheses propose that repeated drug exposure leads to particular changes in neural responses associated with certain stimuli and behavior. Moreover these neuronal changes are expected to be associated with the emergence of addiction-like patterns of behavior. Electrophysiological recordings in behaving animals can be used to test these hypotheses. For example, the incentive sensitization hypothesis postulates that repeated exposure to addictive drugs will lead to increased incentive-related neuronal responses to drug-associated conditioned stimuli in the mesoaccumbal dopamine (DA) pathway. The homeostatic dysregulation hypothesis might predict that repeated drug exposure will cause a lasting shift in activity of NAc neurons that track hedonic state (if such exist) and a decreased effect of natural rewards on those neurons in animals with extensive drug histories. Based on the S-R hypothesis, dorsal striatal neuron responses involved in regulating instrumental behaviors are expected to be amplified after repeated drug exposure; whereas ventral striatal responses involved in flexible behavior would be weakened. Based on the differential neuroplasticity hypothesis one would predict both increased responses to drug-related events (such as drug-associated cues) and decreased responses to natural reward-related events (such as cues associated with natural rewards). In all cases, as the predicted changes in firing emerge, addiction-like behaviors are also expected to develop.

Electrophysiological recordings in behaving animals can be used to test the addiction hypotheses. Specifically, the procedure can be used to (1) identify baseline neural responses that correspond to those implicated by a specific hypothesis; (2) test for predicted effects of repeated drug exposure on those neural responses; and (3) test whether those changes in neuronal responses are associated with the emergence of the predicted, addiction-like behaviors. One of the most challenging aspects of this work is to identify neural responses associated with specific types of information related to motivated behavior. There are several approaches that could be used. First, one can use paradigms that temporally dissociate particular processes and events. In this case, neural responses related to one or more of those processes and events are expected to show a response at some time points and not at others. Alternatively, the paradigm can be designed to manipulate individual processes, holding others constant. If a neural response reflects the manipulated variable, then one should observe predicted changes in that response (i.e., prevalence or magnitude of response).

Sophisticated examples of these approaches can be found in primate recording studies of DA and dopaminoceptive target regions such as the ventral and dorsal striatum (Schultz 2000). These studies aimed to identify the specific role of DA and dopaminoceptive neurons in motivated behavior. Early studies showed that the neurons respond to discriminative stimuli that predict reward availability. Discriminative stimuli carry multiple types of information and activate multiple processes, including but not limited to response–reinforcer associations, which facilitate the occurrence of a specific operant, and stimulus–reward associations, which might activate Pavlovian behaviors and visceral processes related to consumption. Single neuron responses theoretically could have corresponded to one or more of those associations and events. It was also possible that changes in firing time-locked to the discriminative stimulus reflected sensory properties of the cue. One strategy used to differentiate among these possibilities was to temporally dissociate the different types of information. Animals were trained to discriminate multiple cues, which were presented separately. Each cue provided a particular type of information (e.g., the particular operant response that was required to earn the reward during an upcoming trial or the particular reward that was available on a given trial) (Apicella et al. 1991; Cromwell and Schultz 2003). These cues were temporally spaced so that researchers could test for changes in neural responses to each of the cues. Some neurons responded to a single cue, and other neurons responded to multiple cues. With this approach it was possible to differentiate among the types of information to which changes in neuron firing corresponded.

To characterize potential behavioral correlates of firing patterns, researchers manipulated particular variables, holding others constant. For example, the researchers were interested in testing whether firing patterns during instrumental behavior were related to the specific behavior. To test this possibility animals were trained to emit either of two behavioral responses, depending on which of two discriminative stimuli were presented. In these experiments, neural responses were stable across the trials that required different operant behaviors, indicating that those responses were unrelated to the cue-response associations. A similar approach can be used to test whether a cue response reflects primary sensory processing or a response to the associative characteristic of the cue. For example, researchers have trained animals to discriminate cues, which signal different rewards, and then tested the effect of reversing the cue–reward relationship. In such experiments, cue responses that reflect sensory properties of the cue are unaffected by the reversal of the cue–reward relationship. On the other hand, neural responses that reflect the associative (i.e., cue–reward relationship) properties of the cue show changes in activity (Cromwell and Schultz 2003; Calu et al. 2007; Roesch et al. 2007; Stalnaker et al. 2007).

Similar approaches can be used to isolate neuronal responses relevant to particular drug addiction hypotheses. Thus far, recording studies related to addiction have employed some of the simplest possible procedures. This is in part related to the multiple challenges associated with a rat model of intravenous self-administration, including but not limited to the time required to condition rats on sophisticated behavioral procedures and the limited life span of the intravenous catheter. However, advances related to testing addiction hypotheses have been made and more sophisticated studies are possible.

2.5 FIRING PATTERNS DURING COCAINE TAKING

2.5.1 Overview of the Section

One focus of recording studies of cocaine-directed behavior has involved the cataloging of firing patterns that occur during FR1 cocaine self-administration sessions. Another has been investigating the nonpharmacological factors that contribute to those firing patterns. This research is of interest for at least two reasons. First, identifying firing patterns that are nonpharmacological facilitates identification of those firing patterns that are potentially pharmacological. Additionally, as we have already described, the recording method can be used to test current hypotheses regarding NAc mechanisms that contribute to the development and expression of drug addiction. This research requires isolating neural responses that are related to particular motivational events.

Much progress has been made in documenting firing patterns exhibited by NAc neurons in animals with limited drug exposure. Some experiments have also provided information about the events that are encoded by the firing patterns, though more work in this area is necessary. This section of the chapter reviews some of this research, which exemplifies numerous methods that can be used to differentiate among particular nonpharmacological determinants of firing patterns, as well as to differentiate between pharmacological and nonpharmacological underpinnings of the firing patterns.

2.5.2 Short-Duration Changes in Firing Time-Locked to the Reinforced Lever Press

2.5.2.1 Description of the Firing Patterns

Some of the first extracellular recordings conducted during FR1 cocaine self-administration sessions characterized phasic firing patterns time-locked to the operant behavior. These studies (Carelli et al. 1993; Chang et al. 1994; Peoples and West 1996; Uzwiak et al. 1997) showed that at intermediate dose of cocaine (e.g., 0.75 mg/kg/inf), approximately 25%–40% of NAc neurons exhibit rapid phasic changes in firing time-locked to cocaine-reinforced lever presses (referred to herein as lever-press firing patterns). The lever-press firing patterns vary in time course, with the change in firing occurring primarily within a 9-sec period that brackets the reinforced press (i.e., 3-sec prepress to the 6-sec postpress) (Figure 2.6). We have subtyped these firing patterns into the following five groups: (1) exclusively prepress, (2) predominantly prepress, (3) symmetrically pre-post press, (3) predominantly postpress, and (4) exclusively postpress.

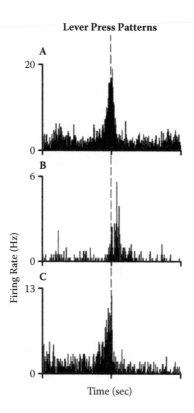

FIGURE 2.6 Rapid phasic changes in firing time-locked to the cocaine-reinforced lever press during an FR1 cocaine self-administration session. Each histogram shows an individual neuron example of an excitatory lever-press firing pattern. In each of the three histograms, average firing rate (Hz per 0.1-sec bin) is plotted during the 12 sec before and after presses made during the maintenance phase of the self-administration session (i.e., 10th–27th reinforced lever presses). Figure taken from Peoples et al., *J Neurophysiology* 91:314–323, 2004. Other subtypes of phasic firing pattern are shown in Uzwiak et al., *Brain Res* 767:363–369, 1997; and Peoples and West, *J Neurosci* 16(10):3459–3473, 1996.

Carelli and colleagues have described a "dual peak" firing pattern time-locked to the cocaine reinforced operant, which they have thus far not observed in studies of nondrug rewards. The pattern consists of two discrete phasic changes in firing, one prepress and one postpress. However, patterns similar to the dual peak response time-locked to the cocaine-reinforced operant have been described in primate studies of natural rewards (Schultz et al. 1992; Schultz 2000). It thus appears that the patterns are not unique to cocaine.

Phasic lever-press firing patterns have been observed under a broad range of experimental conditions but have been consistently reported as predominantly excitatory (Carelli et al. 1993; Chang et al. 1994; Uzwiak et al. 1997; Hollander and Carelli 2005; Peoples et al. 2007b). The firing patterns occur in both core and anterior portions of the shell but are nearly absent in the posterior portions of the shell (Uzwiak et al. 1997).

In addition to these rapid phasic firing patterns a small additional portion of neurons ($\approx$30%) show longer duration (e.g., 30–60 sec) lever-press firing patterns (not shown). These neurons show either an increase prepress, an increase postpress, or a decrease postpress. Though the patterns are long in duration, they are punctuate with rapid onsets and offsets (Peoples and West 1996). The broad range of firing pattern time courses suggests that during cocaine self-administration, NAc neurons track multiple variables or combinations thereof during the seconds that precede and follow the reinforced operant response.

2.5.2.2 Determinants of the Firing Patterns

The offset of the "rapid" lever-press firing patterns precedes the expected latency for pharmacological actions postinfusion. There are numerous nonpharmacological events that occur within the seconds that lead up to and follow the reinforced lever press during typical FR1 cocaine self-administration sessions. These include, but are not limited to the following: approach to the lever, the lever-press operant, the conditioned reinforcer cues, and delivery of the primary reinforcer. The phasic firing patterns time-locked to the operant potentially encode information related to one or more of these events.

Studies were conducted to identify the events associated with the phasic firing patterns. These studies employed a number of analytic and experimental strategies. For example, researchers compared the time course of the firing patterns and the occurrence of specific behaviors during the seconds leading up to and following completion of the lever-press operant (Chang et al. 1994). In other studies, investigators dissociated particular events in time or manipulated particular events, while holding other nonpharmacological and pharmacological variables constant (Peoples et al. 1997; Carelli 2000). We will describe one example of the latter approach in this subsection.

In one study (Peoples et al. 1997), we tested for responses that were related to either the conditioned reinforcer or the instrumental behavior. Animals were trained on the FR1 cocaine self-administration session. We then conducted a multiphase recording session. At the beginning of the session, subjects self-administered cocaine infusions as usual. Thereafter, the session consisted of alternating phases in which infusions were either response contingent or response noncontingent. During each response-contingent phase, presentation of the conditioned cues and the associated cocaine infusion (5 or 15) occurred only when the rat depressed the lever. In each response-noncontingent phase, the presentation of the cue and infusion occurred in the absence of operant behavior. The infusion was presented with the exteroceptive light plus tone cue because proprioceptive or pump-related auditory cues might be associated with the onset of the infusion, which could accrue conditioned stimulus properties. The timing of the cue-infusion presentations followed the schedule of cue + infusion presentations self-administered by the rat during the preceding response-contingent phase. Lever presses during the noncontingent phase were nonreinforced (i.e., were not followed by either cues or cocaine infusions) (Figure 2.7).

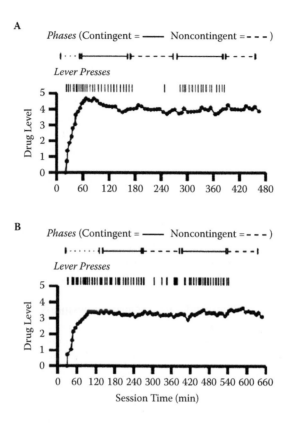

FIGURE 2.7 Patterns of lever presses and calculated drug levels during the response-contingent and response-noncontingent cocaine phases. Patterns of behavior are shown for each of two animals during the successive phases of response-contingent, response-noncontingent, and response-contingent cocaine administration. Within each panel (A and B) is shown the following: each point on the graph shows the calculated drug level (mg/kg) at the time of each single infusion. Drug level is plotted as a function of successive infusions for the initial self-administration (loading) phase, indicated by dotted line, and for the subsequent alternating phases of contingent and noncontingent cocaine infusions. The figure demonstrates the stability of drug levels across response-contingent and response-noncontingent phases, as well as the stability of lever-press behavior between the two response-contingent phases (i.e., the nondisruptive nature of the manipulation). Consistent with the pharmacokinetics of cocaine in the rat (Misra et al. 1976), the drug level was calculated assuming first-order pharmacokinetics using the following equation $(B + D)e^{(-kt)}$ in which B = drug level at the time of previous infusion; D = infusion dose (mg/kg/inf); t = minutes elapsed between the infusion for which drug level is being calculated and the preceding infusion; k = rate constant for cocaine, derived from half-life (i.e., $k = 0.693/t_{1/2}$) of cocaine in the NAc of rats administered a single intravenous injection of cocaine (Hurd et al. 1988). Figure taken from Peoples et al. 1997, *Brain Research* 757:280–284.

The procedure controlled for multiple factors that could potentially influence NAc firing. Given that the schedules of contingent and noncontingent infusions were identical, drug level at the time of the infusion was held constant. Hence, the motivational state of the animal at the time of the cue and infusion was expected to be quite similar between the phases. Given the selective elimination of operant behavior during the seconds preceding the response-noncontingent cue presentations, firing patterns that were associated with the operant were expected to be present during the response-contingent phase but not during the response-noncontingent phase. Such firing patterns were also expected to occur when animals made nonreinforced operant responses. On the other hand, firing patterns that were associated with the conditioned cues were expected to persist across the response-contingent and response-noncontingent phase.

Of the 70 neurons recorded in the study, 29 showed a lever-press firing pattern during the contingent phases of the session. During the noncontingent infusion phase, all neurons (4/4) that showed a predominant prepress firing pattern time-locked to the response-contingent cue presentations did not show a change in firing time-locked to the response-noncontingent cue presentations. The same was true for almost all of the neurons (8/9) that showed a symmetrical firing pattern, and for about half of the neurons that showed either a predominantly (3/7) or exclusively (3/6) postpress firing pattern. For the other neurons (i.e., 10% of symmetrically responsive neurons and half of the predominantly and exclusively postpress responsive neurons), the neurons showed a phasic response time-locked to both response-contingent and response-noncontingent cue presentations, though the responses during the noncontingent phase tended to be diminished relative to that during the contingent phase (Figures 2.8–2.9). These findings show that predominantly and exclusively prepress firing patterns, almost all symmetrical lever-press firing patterns, and half of the predominantly and exclusively postpress firing patterns reflect neuronal responses associated with the operant. For the other neurons, the firing patterns reflect neuronal responses associated with both the operant and the cues. These findings are consistent with those of other studies (Carelli and Ijames 2000; Carelli and Wightman 2004).

Several studies showed that the lever-press firing patterns are specific for the expected reward. In these studies, NAc activity was recorded during multiphase sessions in which the reinforcer varied between phases and was either a natural reward (water or food) or cocaine (Carelli 2000). During these sessions most neurons that showed a phasic response time-locked to cocaine-reinforced lever presses did not show a phasic response time-locked to lever presses reinforced by the natural reward (Carelli et al. 2000). Similarly, other neurons that exhibited responses time-locked to lever presses reinforced by the natural reward rarely did the same when the operant was cocaine reinforced. In some of these experiments, the animals pressed the same lever during both phases, so the between-phase difference in firing could not have reflected either the movements associated with the operant or the spatial location of the manipulanda. These findings are consistent with the interpretation that the lever-press firing patterns are related in some way to expectation of a particular reward.

Additional studies are required to further specify the information encoded by the phasic firing patterns. However, available data on cocaine show that task-related neural activity during cocaine self-administration sessions is quite similar to that

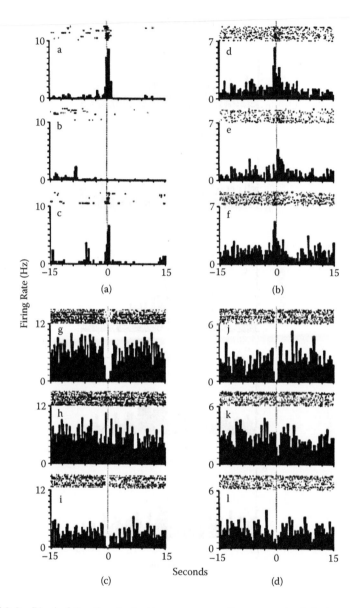

FIGURE 2.8 Phasic firing time-locked to response-contingent cocaine infusions, response-noncontingent cocaine infusions, and unreinforced lever presses. Each panel (A–D) shows the firing patterns of a single neuron. All neurons in the figure exhibited a phasic firing pattern during the first response-contingent phase that had a prepress onset and a postpress offset. In each panel, top-down, firing rate is plotted in relation to the following events (time zero): (1) response-contingent infusions that were presented in conjunction with a tone + light cue (7.0 sec) (histograms a, d, g, j); (2) response-noncontingent infusions presented in conjunction with the tone + light cue (b, e, h, k); and (3) nonreinforced lever presses (i.e., no cocaine infusion and no tone + light cues (c, f, i, l) during the response-noncontingent phase. Figure taken from Peoples et al. 1997, *Brain Research* 757:280–284.

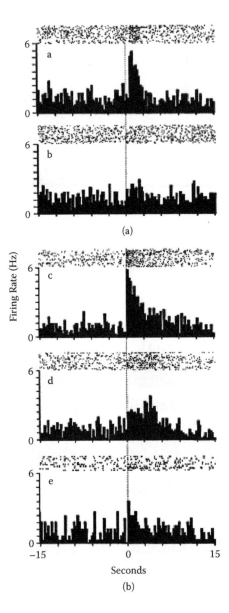

FIGURE 2.9 Phasic firing time-locked to response-contingent cocaine infusions, response-noncontingent cocaine infusions, and unreinforced lever presses. Both panels (A and B) show firing of a single neuron. Both neurons exhibited an exclusively postpress firing pattern during the first response-contingent phase. For each panel, top-down, firing rate is plotted in relation to the following events (Time 0): (1) response-contingent infusions that were presented in conjunction with the tone + light cue (a, c); and (2) response-noncontingent infusions that were presented in conjunction with the tone + light cue (b, d). For Panel B, firing rate is also displayed in relation to nonreinforced lever presses (i.e., no cocaine infusion and no tone + light cue) that were made during the response-noncontingent phase (e). Figure taken from Peoples et al. 1997, *Brain Research* 757:280–284.

which occurs during behavior directed toward natural rewards. In both cases, NAc neurons exhibit phasic firing in association with stimulus events that engender an incentive state (e.g., cues that signal the availability of reward) and to behavioral events that are controlled by those states (Schultz et al. 1992; Lavoie and Mizumori 1994; Bowman et al. 1996; Hollerman et al. 1998; Cromwell and Schultz 2003). The firing patterns are exhibited by a relatively small subset of recorded neurons. The patterns involve predominantly increases in firing. The excitations are dissociable from both specific movements and physical properties of the cues and depend on reward expectation. The nature of the relation to reward expectation has been investigated in some detail in studies of natural rewards. Those studies show that the cue-associated firing patterns depend strongly on the predicted reward. Specifically, the magnitude of cue-locked phasic firing patterns varies depending on the following: (1) the reward type predicted by the cue (Hassani et al. 2001), (2) the magnitude of the predicted reward (Hollerman et al. 1998; Cromwell and Schultz 2003), and (3) the temporal proximity of the reward (Shidara et al. 1998). Whether the same is true for the cue-associated firing patterns during cocaine self-administration sessions remains to be determined.

Studies of natural rewards have also shown that NAc neurons exhibit changes in firing during reward consumption (Schultz 2000; Nicola et al. 2004). In contrast to the phasic activity time-locked to reward-predictive cues and operant behavior, the majority of the changes in firing during reward consumption are inhibitory. Though cocaine self-administration is not associated with specific consummatory behaviors, a high percentage (50%) of neurons show an inhibition in firing during the min(s) following each cocaine infusion. The nature of these decreases is not yet understood; however, it is possible that the change in firing is analogous to the inhibitory response exhibited by NAc neurons during consumption of natural rewards.

2.5.3 LONG-DURATION FIRING PATTERNS TIME-LOCKED TO THE COCAINE-REINFORCED LEVER PRESS: THE PROGRESSIVE REVERSAL FIRING PATTERN

2.5.3.1 Description of the Firing Pattern

NAc neurons exhibit another category of firing pattern time-locked to the cocaine-reinforced lever press (Peoples et al. 1994, 1998b; Peoples and West 1996). At an intermediate dose of cocaine (0.75 mg/kg/inf), approximately 65% of all neurons show a change in firing that occurs slowly across the interval between successive self-infusions. For most neurons (i.e., 50% of all recorded neurons), firing rate decreases during the first minute postpress and then progressively increases until the time of the next lever press (i.e., decrease + progressive reversal) (Figure 2.10). For the remaining neurons, firing rate increases postpress and progressively decreases until the time of the next press (i.e., increase + progressive reversal) (Figure 2.10).

2.5.3.2 Determinants of the Firing Patterns

Investigation of the progressive reversal firing pattern exemplifies some of the methods that can be used to differentiate between nonpharmacological and pharmacological determinants of firing patterns observed during periods of drug exposure. In

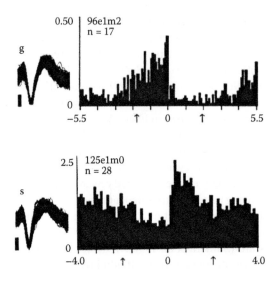

FIGURE 2.10   The progressive reversal firing patterns time-locked to the cocaine-reinforced lever press during an FR1 cocaine self-administration session. The figure shows single-neuron examples of a decrease + progressive reversal firing pattern (top) and an increase + progressive reversal firing pattern (bottom). In each histogram, average firing rate (Hz per 0.6-sec bin) is plotted during minutes pre- and postpress. The time base of each histogram corresponds to the average interinfusion interval exhibited by the individual animal from which the neural activity was recorded. Figure taken from Peoples et al., *J Neurosci* 18(18):7588–7598, 1998.

particular, the research demonstrates use of the following: (1) time course comparisons of firing patterns and pharmacological events, (2) time course comparisons of firing patterns and nonpharmacological events, and (3) the behavioral clamp method. The research also highlights the types of nonpharmacological variables that potentially contribute to NAc firing patterns during drug self-administration.

The decrease + progressive reversal firing pattern parallels changes in locomotion and stereotypy that occur over the course of the interinfusion interval (i.e., between successive self-infusions) (Figure 2.11). Peoples et al. (1998b) tested whether the decrease + progressive reversal firing pattern reflected changes in the frequency of locomotion. To test this interpretation we determined whether neurons that show the decrease + progressive reversal firing pattern also showed phasic changes in firing time-locked to specific locomotion events. Moreover, we used the clamping procedure to test whether the firing pattern depended on the variations in locomotion over the course of the interinfusion interval.

We carried out these tests with the aid of video analysis. During each recording session, behavior was videotaped. Each video frame (33-msec) was sequentially time-stamped by a computer coupled with a video frame counter. Frames were time-stamped according to the same computer clock that time-stamped each neural discharge. After the recording sessions, in offline frame-by-frame analysis, time stamps associated with the onsets and offsets of particular behaviors (33-msec temporal

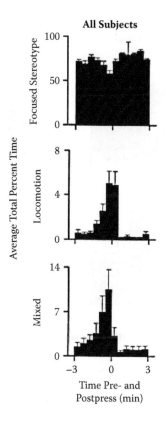

FIGURE 2.11   Average percentage of time spent in locomotion and stereotypy exhibited by all subjects during an FR1 cocaine self-administration session. Average time spent (per 30-sec bin) is plotted as a function of min pre- and postpress. Time 0 = time of reinforced operant response. Figure taken from Peoples et al., *J Neurosci* 18(18):7588–7598, 1998.

resolution) were compiled and used in subsequent histogram analysis of neural firing in relation to the specific behaviors.

Comparisons of average firing during locomotion showed that half of the decrease + progressive reversal neurons showed a significant increase in average firing during locomotion relative to the seconds immediately preceding the onset of the behavior. For almost all of these neurons, this phasic response was selective for either locomotion toward the lever or locomotion away from the lever. The absence of phasic firing for half the neurons and the directional specificity of the phasic locomotion responses for the other neurons showed that the decrease + progressive reversal firing patterns are not related to locomotion per se. The clamping control provided additional evidence against the motor interpretation. Specifically, when histograms displaying firing during the minute before and after cocaine infusion were reconstructed with all periods of locomotion excluded from the calculation of firing rate, none of the decrease + progressive reversal patterns were lost (Figure 2.12). Thus, none of the

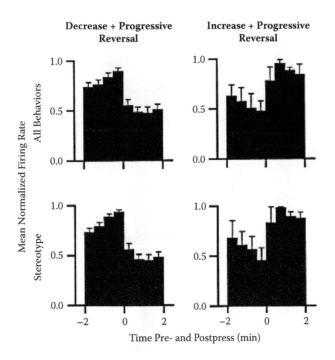

FIGURE 2.12 The change + progressive reversal firing patterns persisted when all locomotion was excluded from calculations of firing rate. Each histogram shows phasic firing time-locked to the cocaine-reinforced lever press. Histograms shown in the left column correspond to all decrease + progressive reversal neurons combined (n=15). Histograms in the right column correspond to all increase + progressive reversal neurons (n=3). For each column, the top histogram displays firing during all behaviors, and the bottom histogram displays firing of the same neurons exclusively during stereotypy. Each histogram displays average normalized firing rate (i.e., Hz per 30-sec bin) during the min before and after the press (ordinate). Individual neuron firing rates were normalized before the group average was calculated by dividing hertz in each 0.5-min bin by the hertz in the 0.5-min bin showing the maximum firing rate in that histogram. Figure 11 from Peoples et al., *J Neurosci* 18(18):7588–7598, 1998.

decrease + progressive reversal firing patterns could be attributed to variations in the frequency of locomotion across the minutes of the interinfusion interval.

Though the firing patterns are not related to movement per se, the above findings show that some of the decrease + progressive reversal firing patterns may be related to the directional, spatial characteristics of locomotion. Alternatively, the patterns, like lever-press firing patterns, might be related to reward expectations. This interpretation is consistent with evidence that mesolimbic DA and the NAc do not directly mediate the execution of movements; instead they have a psychomotor or motivational function that facilitates incentive-related approach and preparatory behaviors (Iversen and Koob 1977; Robbins and Everitt 1982; Beninger 1983; Kelley and Stinus 1985; Wise and Bozarth 1987; Cador et al. 1989, 1991; Everitt et al. 1989; Apicella et al. 1991; Di Chiara et al. 1992; Salomone 1992; Schultz et al. 1993).

In considering possible reward-related determinants of the decrease + progressive reversal firing pattern, it is relevant to note that the firing pattern mirrors changes in drug and DA levels during the cocaine self-administration session (Wise et al. 1995; Peoples and West 1996). The changes in drug level that occur between successive self-infusions are expected to engender interoceptive cues (Colpaert 1987; Overton 1987; Stewart and deWit 1987), including those that develop progressively over the course of the interval, in conjunction with drug metabolism. These interoceptive cues could include or become associated with changes in the motivational state of the animal that occur with changes in drug level. It is possible that the progressive increase in firing reflects excitatory afferent input related to the cues and motivational state of the animal and that the rapid decrease in firing that occurs postpress reflects the cessation of that afferent input. An alternative, but not mutually exclusive, interpretation of the firing pattern is that it reflects more direct acute actions of cocaine in the NAc. The relative contribution of pharmacological and nonpharmacological variables to the decrease + progressive reversal firing patterns remains to be further delineated. However, there is evidence that the patterns cannot be wholly pharmacological. For example, for some neurons the postpress decrease begins before the onset of the cocaine infusion (Peoples and West 1996; Peoples et al. 1999b).

Regardless of the relative contribution of pharmacological and nonpharmacological factors to the firing pattern, we hypothesize that the pattern contributes to regulation of the self-administration behavior. This hypothesis is based on the following observations. As already noted, the pattern closely mirrors changes in drug and accumbal DA levels. Previous research has shown that successive self-infusions occur predictably, at time points when drug and accumbal DA level falls to a particular level. Single trial analysis of the firing rates of decrease + progressive reversal neurons shows that the timing of self-infusions is also associated with the neurons' achieving a fairly reliable apical firing rate. Additional analyses showed that the duration of the progressive increase in firing is positively and significantly correlated with the interinfusion interval: The longer the interinfusion interval, the longer the duration of the progressive increase in firing. The decrease + progressive reversal firing pattern also parallels changes in the propensity of animals to approach the response lever. Based on these observations and the findings of the locomotion and behavioral clamp analyses, the decrease + progressive reversal firing pattern likely influences the propensity of animals to initiate drug-directed behavior.

2.5.4 SESSION-LONG CHANGES IN AVERAGE FIRING RATE

2.5.4.1 Description of the Firing Patterns

Extracellular recordings during intravenous FR1 cocaine self-administration (0.75 mg/kg/inf) sessions have shown that most ($\approx$90%) NAc neurons exhibit a change in average firing rate during the self-administration session relative to the pre- and post-session baseline periods (referred to herein as session-long increases and decreases, Figure 2.13) (Chang et al. 1998; Peoples et al. 1998a, 2004, 2007b). The session-long changes in firing are exhibited by neurons that exhibit a phasic lever-press firing pattern, a progressive reversal firing pattern, or both. But, the

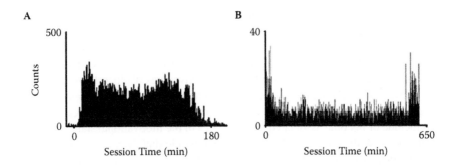

FIGURE 2.13 Session-long changes in firing during FR1 cocaine self-administration sessions. The figure shows single-neuron examples of a session-long increase (left) and a session-long decrease (right) in firing. In each histogram, average firing (Hz per 30-sec bin) is plotted during an entire FR1 cocaine self-administration session, as well as during a pre- and postsession baseline period.

patterns are also exhibited by neurons that show no other change in firing (Peoples et al. 1998b). At the dose of 0.75 mg/kg/inf, the majority of the session-long changes in firing (i.e., 60% percent of all recorded neurons) are decreases.

2.5.4.2 Determinants of the Firing Pattern

The session-long change in firing pattern is defined by a sustained change in average firing during the self-administration session relative to the drug-free baseline period. It thus involves a difference in firing between a drug-free period and the entire period of drug exposure and potentially reflects an acute drug effect. Though the firing pattern has a time course consistent with a potential pharmacological origin, it is also possible that the firing patterns reflect NAc encoding of nonpharmacological events. The self-administration session and the baseline periods differ in a number of ways. During the baseline period animals generally remain at rest, locomoting only occasionally. However, during the self-administration session animals engage in operant behavior and exhibit increased locomotion and drug-induced stereotypic behaviors. The session-long changes therefore potentially reflect pharmacological actions of cocaine, afferent input associated with the behaviors that are unique to the self-administration session, or both.

A number of studies have been conducted to differentiate among the various interpretations of the session-long change in firing. The investigations provide examples of numerous methods that can be employed to differentiate pharmacological and nonpharmacological firing patterns. These include the following: (1) time course comparisons between the firing patterns and pharmacological events, (2) time course comparisons between the firing patterns and nonpharmacological events, (3) comparisons of firing patterns during drug-directed behavior and similar natural reward-directed behavior, (4) behavioral clamp procedures, and (5) comparisons with natural reward. The description of the firing patterns is presented in two sections of this chapter. Here we limit the discussion to research that demonstrates a nonpharmacological component to the firing patterns. Research related to pharmacological determinants of the firing patterns is overviewed in Section 2.7.

Two lines of evidence are consistent with the hypothesis that at least some portion of the session-long changes in firing is nonpharmacological. First, the onset of some of the changes in firing precedes the delivery of the first drug infusion and occurs in association with the cues that signal the onset of drug availability and the onset of the first operant response (Peoples et al. 2004). Second, NAc neurons exhibit session-long changes in firing during FR1 sucrose-reinforced operant behavior (Kravitz et al. 2006; Kravitz and Peoples 2008). These observations are consistent with the hypothesis that at least some of the session-long changes in firing reflect nonpharmacological processes.

In considering possible nonpharmacological determinants of the firing patterns, we first tested the hypothesis that the patterns reflect an effect of phasic firing time-locked to the operant behavior on average firing. Several lines of evidence led to the rejection of this hypothesis. First, the majority of session-long changes in firing do not exhibit a phasic change in firing time-locked to the lever-press operant. Second, in 75% of the cases in which phasic and session-long changes co-occur, the changes are directionally opposed (Peoples et al. 2004; unpublished observations). Third, session-long changes in firing are apparent regardless of whether firing rate calculations are limited to either periods in which the animal is engaged in drug-directed behavior, or periods in which other behaviors such as stereotypy occur (Peoples et al. 2004 and Figure 2.14). These observations are consistent with the interpretation that

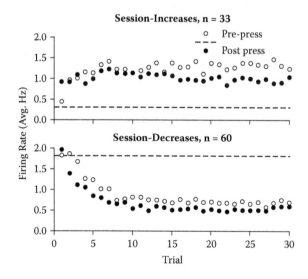

FIGURE 2.14   Average firing rates of session-increase (top) and session-decrease (bottom) neurons during specific 30-sec periods before and after each lever press of a single cocaine self-administration session. Firing is shown during the -120 to -90 sec before each press and +90 to +120 sec after each press. Video analysis shows that animals engage almost exclusively in stereotypy during these periods. Average firing rates during the 30-sec periods are plotted as a function of press number during the self-administration session. Dashed horizontal lines reflect average presession baseline firing rates of the session-increase neurons (top graph) and the session-decrease neurons (bottom graph).

session-long changes in firing do not reflect an effect of phasic firing time-locked to operant behavior, stereotypy, or locomotion.

2.5.5 FIRING PATTERNS DURING COCAINE TAKING: SECTION SUMMARY, NOVEL FINDINGS, AND RESEARCH OPPORTUNITIES

One goal of experiments that employ the chronic microwire recording method is to identify acute pharmacological effects on NAc neural activity, which might contribute to cocaine addiction. In behaving animals, a change in single neuron activity that occurs with a change in drug exposure can reflect the following: (1) pharmacological actions on the recorded neurons or primary afferent inputs, or (2) neural responses to differences in behavior, stimulus events, or other factors between the drug-exposed and drug-free condition. For behaving animal recordings to be useful in identifying pharmacological mechanisms that contribute to addiction, it is necessary for those studies to differentiate the relative contribution of the pharmacological and nonpharmacological variables to changes in firing patterns associated with drug exposure. A number of strategies can be employed to make this differentiation. The research reviewed in this section exemplifies a number of them. Additional approaches are described in later sections of the chapter. The research also identified two firing patterns that potentially reflect acute actions of self-administered cocaine. Subsequent sections describe additional chronic microwire recording studies of the pharmacological effects of cocaine on NAc neurons.

An additional potential application of the chronic microwire recording method is to investigate drug-induced changes in information processing that may underlie the abnormal motivated behavior that characterizes addiction. It is possible that the development of the behaviors is mediated in part by acute drug-induced alterations in information processing, which influence normal adaptive processes associated with learning and memory. If this were the case one might expect to observe clear differences in event-related NAc neural responses during operant behavior that occurs in the presence versus absence of cocaine.

Available evidence shows that there are both similarities and differences between task-related neural activity during cocaine- and natural reward-directed instrumental behavior. Some of the similarities include the predominance of excitatory changes in firing time-locked to predictive cues and instrumental behavior, and evidence that the firing patters are related to reward-related expectations. Thus far, the most prominent differences include the greater prevalence of session-long decreases in firing during drug-reinforced operant sessions, the progressive reversal firing pattern, and the dissociation between neurons that show task-related firing during behavior directed toward cocaine and natural rewards. The functional significance of these differences is not fully understood. Further evaluation of the question has implications for understanding the role of acute drug actions in the development of drug addiction. Strong overlap would suggest that acute drug-induced changes in information processing play a limited role; whereas significant differences suggest that those drug actions might play an important role in the development of addiction.

2.6 FIRING PATTERNS DURING COCAINE SEEKING

2.6.1 Overview of Section

In addition to characterizing firing patterns during FR1 cocaine self-administration sessions, researchers have begun to document and investigate firing patterns during periods in which animals engage in drug-directed behavior under drug-free conditions. There are a number of reasons for conducting these types of studies. First, neuropharmacological studies show that the mechanisms that mediate drug-directed behavior under drug-free and drug-exposed conditions are not identical (Whitelaw et al. 1996; Arroyo et al. 1998; Markou et al. 1999; Grimm and See 2000; Olmstead et al. 2000, 2001; Everitt et al. 2001; Cardinal et al. 2002; Dickinson et al. 2002). Second, drug-directed behavior under drug-free conditions (referred to as drug-seeking) is thought to better model conditions associated with relapse in humans, which is a primary therapeutic target (O'Brien 2005). Third, the research also represents an additional approach to differentiating firing patterns that are mediated by acute drug actions from those that are nonpharmacological. NAc firing patterns associated with drug-directed behavior in drug-free animals have been characterized in three basic types of experimental situations. The experiments and findings are described in this section with the intent of providing further overview of methods but also to highlight novel findings and a unique research opportunity afforded by the chronic-behaving animal recording procedure.

2.6.2 Experiments

2.6.2.1 NAc Firing at the Onset of an FR1 Cocaine Self-Administration Session

In one study we characterized NAc responses during cue-evoked drug-directed behavior in drug-free animals. Recordings were made in animals with a two-week history of daily cocaine (0.75 mg/kg/inf) self-administration (Peoples and Cavanaugh 2003; Peoples et al. 2004). The recording session consisted of a 20-min presession baseline period, a cocaine self-administration session, and a 40-min postsession baseline period. During the presession baseline and postsession recovery period, animals did not have access to the operant response lever, and the drug was not available. The onset of the cocaine self-administration was signaled by presentation of a compound tone (7.5 sec) + light (30 sec) cue. The cue was the same as that typically associated with presentation of the primary cocaine reinforcer. Animals initiated cocaine self-administration shortly after the onset of the cue, but not within less than 1 min.

Average firing during the 30-sec cue that signaled the onset of the self-administration session (referred to as the discriminative stimulus, S^D) was compared with firing during the 30-sec preceding the cue. Changes in firing during the 30-sec cue presentation relative to the 30-sec precue period were referred to as S^D changes in firing. We additionally compared average firing during the 30 sec that preceded the first cocaine-reinforced operant response and the 30-sec pre-S^D. Changes in firing during this period were referred to as pre-first-press responses.

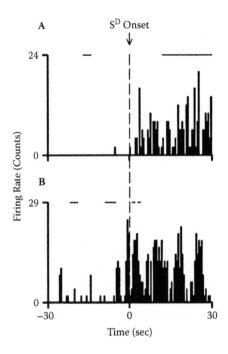

FIGURE 2.15   Excitatory changes in firing time-locked to presentation of cues that signaled the onset of an FR1 cocaine self-administration session. The change in firing time-locked to the onset of a cue that signaled the start of a cocaine self-administration session is shown for each of 2 neurons (1 per row) (firing pattern referred to as an S^D response). Firing rate (counts per 0.5-sec bin) is plotted for the 30 sec before and after onset of the cue. Cue onset is shown at Time 0 on the abscissa. Above each of the histograms are horizontal lines indicating periods of locomotion. Comparison of the timing of locomotion and the time course of the change in NAc firing shows that the latter is dissociated from, and thus not attributable to, the former. Figure taken from Peoples et al., *J Neurophysiology* 91:314–323, 2004.

The findings of this study showed that about 20% of all recorded NAc neurons respond during presentation of the cue. About 60% of the cue-responsive neurons also show a first-press response. An additional 10% of all recorded neurons show a first-press firing pattern without showing an S^D response. Almost all (>85%) of the S^D and pre-first-press responses are excitatory (Figures 2.15–2.16). In many cases, S^D responses are sustained during the minutes that elapse between the onset of the S^D and the occurrence of the first press. The sustained nature of the S^D firing pattern is consistent with the interpretation that for a majority of neurons; the S^D response is not related to the physical properties of the cue. The timing of the firing patterns is also dissociable from that of specific behaviors during the cue and also during the 30-sec pre-first press (e.g., Figure 2.15). The total percent of neurons that show either an S^D or first-press response is consistent with the prevalence of neurons that show lever-press firing patterns during the cocaine self-administration session. Moreover, the majority of neurons that show either an S^D or pre-first-press firing

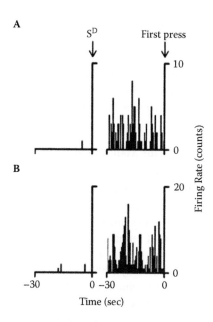

FIGURE 2.16 Excitatory changes in firing associated with the onset of drug-directed behavior during an FR1 cocaine self-administration session. The figure shows two single-neuron examples of changes in firing time-locked to the 1st cocaine-reinforced press (i.e., pre-1st-press responses). Each row of the figure shows firing exhibited by the same single neuron. In each histogram, firing rate (counts per 0.5-sec bin) is plotted for the 30-sec pre-S^D (left) and the 30-sec pre-1st press (right). Figure taken from Peoples et al., *J Neurophysiology* 91:314–323, 2004.

pattern also show a phasic increase in firing time-locked to the cocaine-reinforced lever press.

2.6.2.2 NAc Firing Patterns during Cue-Evoked and Nonreinforced Drug-Directed Behavior

In another study (Ghitza et al. 2003), NAc recordings were conducted during periods in which animals were first presented with a drug-associated cue and then initiated a period of unreinforced drug-directed operant behavior (i.e., a period of extinction). In the study, Ghitza et al. (2003) trained rats on a tone discrimination paradigm. Specifically, during the self-administration sessions, individual lever presses that occurred during tone presentation produced an intravenous infusion of cocaine (0.35 mg/kg/inf) and terminated the tone. Presses made in the absence of the tone were nonreinforced. Once animals showed stable operant responding, animals were exposed to a 3–4 week period of abstinence. Animals were then returned to the operant chamber and exposed to a single presentation of the cocaine-predictive tone. NAc single-unit activity was recorded during re-exposure to the tone under extinction conditions (i.e., operant behavior was not reinforced and animals remained drug free).

Of the neurons recorded during extinction, 37 neurons (53%) exhibited at least a twofold, tone-evoked change in firing within 150 msec after tone onset, relative to baseline firing during the 150-msec time period preceding tone onset. Of the NAc neurons exhibiting tone-evoked activity, 76% were excited and 24% were inhibited by the tone. Although the tone-evoked change in firing of these neurons commenced before the onset of drug-directed behavior, the change in firing persisted after the onset of the behavior for 86% of the neurons. Control analyses showed that the changes in firing were not attributable to motor behavior and did not reflect a primary sensory response.

2.6.2.3   Drug-Directed Behavior Maintained by Presentation of a Conditioned Reinforcer

Studies of NAc dopamine (DA) show that the accumbal response to the presentation of unexpected, experimenter-delivered cue presentations is not the same as that associated with presentation of expected, response-conditioned presentations (Ito et al. 2000, 2002). The behavior of humans is thought to be partially directed and maintained by positive conditioned predictors and conditioned reinforcers (O'Brien et al. 1990; Childress et al. 1993; Tiffany 1999; Garavan et al. 2000). Based on these observations, it may be important to characterize NAc neural activity in relation to drug-directed behavior maintained by conditioned reinforcers. We have begun to approach this issue using a multi-phase seeking-taking task.

The multi-phase task is conducted as follows. During the first phase of the two-phase task, animals engage in operant behavior reinforced on an FR10(FR10:S) schedule of reinforcement (one to five trials). Under this second-order schedule, animals earn 10 conditioned reinforcer (i.e., S) presentations on an FR10 schedule of reinforcement. On this schedule of conditioned reinforcement, every 10th press was reinforced by presentation of a tone + light cue (i.e., the conditioned reinforcer). The 10th presentation of the conditioned reinforcer is paired with a cocaine infusion. Thus, each cocaine infusion is preceded by the completion of 100 operant responses. Animals earn five reinforcers on this schedule of reinforcement during the first, "seeking" phase of the session. During the second phase of the session, an FR1 TO 60 sec schedule was in effect. This schedule establishes a substantive period of drug-free drug-directed behavior (i.e., during the first trial of the session). This procedure is meant to mimic periods of drug-free drug-directed behavior that might precede a period of drug taking in humans.

In one experiment, we trained four rats to self-administer cocaine (0.75 mg/kg/inf) using the multiphase task. Analysis of NAc firing showed that a subset of neurons exhibited a sustained change in average firing during the first, drug-free trial of the self-administration session. For some neurons (about 30% of recorded neurons) the change was an increase, but an equal number of neurons showed a decrease in average firing (Figure 2.17). In addition to the sustained changes in firing, we observed that a subset of neurons exhibited phasic changes in firing time-locked to delivery of the conditioned reinforcer (Figure 2.18). These changes in firing were exclusively excitatory. The neurons that showed a phasic response time-locked to the conditioned reinforcer also showed an average excitatory response time-locked to the cocaine-reinforced operant during the subsequent "taking" phase of the experiment. Though no phasic decreases in firing occurred time-locked to the conditioned

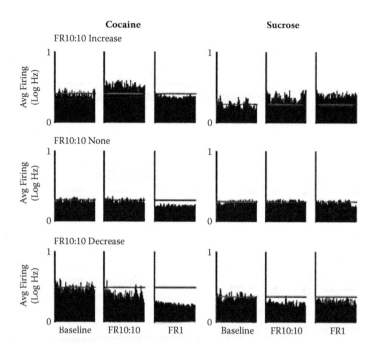

FIGURE 2.17 Average firing of NAc neurons during cocaine- and sucrose-directed operant behavior reinforced by presentation of a conditioned reinforcer. The figure shows average firing rates for three groups of neurons recorded during multiphase seeking–taking operant sessions. The histograms on the left and right of the figure correspond to a cocaine and sucrose session, respectively. The three groups of neurons are the following: (1) neurons that showed a sustained increase in firing during the first trial of the FR10(FR10:S) phase; (2) neurons that showed no sustained change in firing during the first trial of the FR10(FR10:S) phase; and (3) neurons that showed a sustained decrease in average firing during the first trial of the FR10(FR10:S) phase. For each group of neurons, average firing rate is shown for the presession baseline period, the first FR10(FR10:S) trial, and the FR1 taking phase of the session (i.e., operant behavior reinforced on a simple FR1 schedule of primary reinforcement). Taken from Peoples LL, Kravitz AV, Lynch KG, Cavanaugh DJ (2007b) Accumbal neurons that are activated during cocaine self-administration are spared from inhibitory effects of repeated cocaine self-administration. *Neuropsychopharmacology* 32:1141–1158.

reinforcer, there were a small number of neurons that showed such a response time-locked to the cocaine-reinforced operant during the taking phase. In a control study, we trained four additional animals to self-administer sucrose during the two-phase seeking–taking task. Firing patterns during the first, seeking phase of the session were comparable for the sucrose- and cocaine-trained animals (Figures 2.19–2.20).

2.6.3 FIRING PATTERNS DURING COCAINE SEEKING: SUMMARY, NOVEL FINDINGS, AND RESEARCH OPPORTUNITIES

Drug-directed behavior that occurs in the absence and presence of drug is referred to as drug seeking and drug taking, respectively. Various lines of work indicate that the

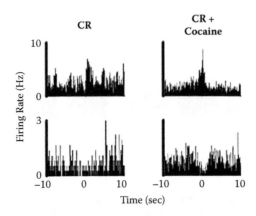

FIGURE 2.18 Group-mean lever-press firing patterns during periods of conditioned reinforcement and primary cocaine reinforcement. Average firing is shown for two groups of neurons recorded during a multiphase seeking–taking cocaine self-administration session. The
groups include the following: (1) neurons that showed a significant phasic increase in firing
time-locked to cocaine-reinforced lever presses during the FR1 taking phase of the cocaine
seeking–taking task (top row); and (2) neurons that showed a significant phasic decrease in
firing time-locked to the cocaine-reinforced lever presses (bottom row). On the left, histograms show average firing time-locked to completion of lever presses reinforced by delivery
of the conditioned reinforcer during the first trial of the FR10(FR10:S) phase of the session.
Time 0 on the abscissa of the histograms corresponds to the concurrent completion of the
operant and the onset of the conditioned reinforcer. On the right, histograms show average
firing time-locked to completion of operant responses reinforced by the combined presentation of the conditioned reinforcer and cocaine.

mechanisms that mediate drug seeking and taking are not the same. Drug seeking is
generally viewed as more relevant to relapse, whereas drug taking is more relevant
to pharmacological events that mediate the development of addiction. The dissociation is viewed as important to successful investigation of the causes and treatment
of addiction.

In this section we reviewed three experiments that characterized NAc activity
during cocaine seeking. Two of the three studies characterized NAc responses to an
initial unpredicted presentation of a drug-associated cue and the subsequent interval
leading up to the first operant (Peoples et al. 2007a). One of these studies additionally
characterized operant responding during a period of extinction. Despite the multiple
parametric differences among the three studies, the findings were quite consistent in
showing that the NAc responses associated with the presentation of drug-associated
cues and the onset of drug-directed behavior are predominantly excitatory and of a
sustained nature. A different profile of NAc firing was observed in animals engaged
in drug seeking maintained by conditioned reinforcement. In particular, there was
a much higher prevalence of sustained decreases in average firing during the period
of drug-free drug-directed behavior. The greater mix of sustained increases and
decreases, as well as the stability of phasic firing patterns time-locked to operant
behavior during the first drug-free trial of conditioned reinforcement and the primary reinforcement (FR1) phase, suggests that patterns of NAc neural activity share

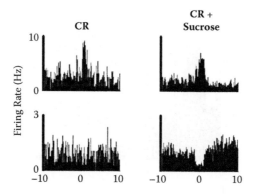

FIGURE 2.19 Group-mean phasic firing patterns during a multiphase surcrose seeking–taking task. Average firing is shown for two groups of neurons recorded during a multiphase seeking–taking cocaine self-administration session. The groups include the following: (1) neurons that showed a significant phasic increase in firing time-locked to sucrose-reinforced lever presses during the FR1 taking phase of the sucrose seeking–taking task (top row); and (2) neurons that showed a significant phasic decrease in firing during time-locked to the sucrose-reinforced lever presses (bottom row). On the left, histograms show average firing time-locked to completion of lever presses reinforced by delivery of the conditioned reinforcer during the first trial of the FR10(FR10:S) phase of the session. Time 0 on the abscissa of the histograms corresponds to the concurrent completion of the operant and the onset of the conditioned reinforcer. On the right, histograms show average firing time-locked to completion of operant responses reinforced by the combined presentation of the conditioned reinforcer and sucrose.

more similarities between drug-free and drug-exposed periods when drug seeking is maintained by conditioned reinforcement than when the behavior occurs during extinction conditions.

The studies of drug seeking have potential implications for investigations that aim to identify causes of relapse and to screen for potential preventative medicines. Most animal studies of neurobiological determinants and potential medications for relapse employ extinction–reinstatement models. Some researchers have observed that the presence versus absence of a history of extinction can alter the neuronal circuits that are involved in reinstatement of drug-directed behavior (Fuchs et al. 2006). The present findings show that NAc neural activity is different during extinction versus conditioned, reinforced drug-directed behavior. Based on these findings, it would seem important to expand studies of relapse to include procedures that evaluate drug seeking under nonextinction conditions.

2.7 ACUTE PHARMACOLOGICAL EFFECTS OF SELF-ADMINISTERED COCAINE

2.7.1 BACKGROUND AND OVERVIEW

Electrophysiological recordings in slice and anesthetized animal preparations show that experimenter-delivered cocaine inhibits spontaneous and glutamate-

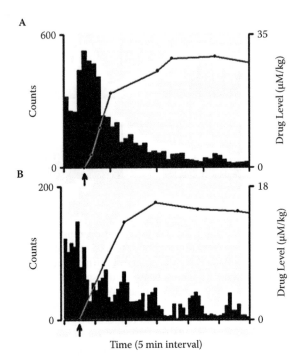

FIGURE 2.20 Average firing of session-long decrease neurons decreases progressively across the initial cocaine self-infusions of the session. Firing of specific accumbal neurons decreases as cumulative drug intake increases during the loading phase of cocaine self-administrations. Each graph plots firing (discharges per 0.5-min bin) plotted as a function of time in the session. Calculated accumbal levels of cocaine are also plotted. Drug levels calculated according to formulas described by Pettit and Justice (1991). Figure taken from Peoples et al., *J Neurophysiology*, 2004, 91:314–323.

evoked firing of NAc neurons (Qiao et al. 1990; Uchimura and North 1990; White et al. 1993; Nicola et al. 1996). This inhibitory effect is mediated primarily by cocaine-induced elevations of accumbal DA and associated increases in activation of DA receptors (for review, see Nicola et al. 2000; also see White et al. 1993, 1998; Hu and White 1994; Henry and White 1995). Ongoing efforts in our laboratory are designed to test whether the acute pharmacological actions of cocaine are also inhibitory.

In testing this possibility we have focused on the session-long changes in firing, which reflect changes in firing during drug exposure relative to pre- and post-session drug-free periods. As already described (2.5.4.2) both the time course of some of the session-long changes and the presence of similar types of changes during sucrose self-administration sessions indicate that normal afferent input associated with reward-directed behavior contribute to the session-long changes in firing. However, the session-long inhibitions tend to be more prevalent during drug self-administrations as compared to natural reward self-administration sessions. It has thus been hypothesized that the session-long changes in firing during cocaine self-administration sessions additionally reflect pharmacological actions of cocaine.' Published and preliminary findings of these efforts are described in this section. The reviewed research demonstrates a number of methods that can be used in identifying pharmacological effects of self-administered drug, including (1) time course comparisons of firing patterns and changes in drug level; (2) dose-response studies; and (3) comparisons to natural reward-directed behavior. The research also demonstrates the importance of extending electrophysiological studies of pharmacological cocaine effects beyond slice and anesthetized animal procedures to include recordings in behaving animals.

2.7.2 SESSION-LONG DECREASES BUT NOT SESSION-LONG INCREASES IN FIRING COVARY WITH DRUG LEVEL

We have made two observations consistent with the interpretation that session-long decreases but not session-long increases are dose-dependent. First, at the beginning of a self-administration session, drug level rises progressively across the first few infusions and then attains a level that is maintained for the duration of the session. Consistent with a possible relationship between session-long inhibition and drug level, onset of the firing pattern, on average, tends to be progressive: firing decreases across the first several infusions of cocaine and then attains a minimum that is maintained for the rest of the session (Figures 2.14, 2.20) (Peoples et al. 1998b, 1999b, 2007b).

Second, a preliminary dose-response study showed that NAc inhibition but not excitation increases with dose of self-administered cocaine. We trained two animals to self-administer cocaine on an FR1 schedule of cocaine reinforcement (0.75 mg/kg/inf) and then exposed each animal to two test sessions. In the first session, a within-session dose-response curve was determined by incrementing the dose of self-administered cocaine according to the following schedule: 0.187, 0.375, 0.75, and 1.5 mg/kg/inf. During the second test session animals self-administered the same doses in a descending order.

A total of 12 NAc neurons were recorded during the two sessions. Animals self-administered 15 infusions of each cocaine dose. To characterize the effect of dose on firing, we calculated firing rate of each neuron during the period that lapsed between the 6th and the 15th infusions and compared these average firing rates to average firing during the presession baseline period (Mann-Whitney tests, $\alpha = 0.05$). Between-dose comparisons showed that both the prevalence and magnitude of inhibitory responses were positively related to dose of self-administered cocaine. There was no effect of incrementing the cocaine dose on the excitatory changes in average firing (Figure 2.21). The positive relationship between drug level and the prevalence and magnitude of inhibitory changes in average firing is consistent with a potential contribution of drug actions to the firing patterns. However, changes in drug level are associated with changes in reward magnitude (Spear and Katz 1991; Thomsen and Caine 2006). It is thus necessary to consider the possibility that dose-dependent changes in the session-long decreases reflect NAc encoding of changes in reward rather than a neurophysiological action of cocaine.

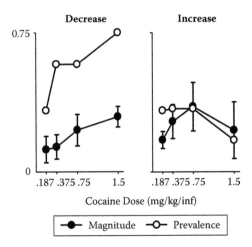

FIGURE 2.21 Changes in average firing of NAc neurons during a within-session dose-response curve. Animals trained to self-administer cocaine (FR1 TO 60 sec; 0.75 mg/kg) were exposed to one of three doses (0.75, 1.5, or 3.0 mg/kg/inf). On the left, prevalence of session-decrease and session-increase firing patterns is plotted as a function of infusion type. On the right, average magnitude of change in firing rate during the 2–6 min postinfusion, relative to baseline. Index is defined as $|(A\text{-}B)|/(A\text{+}B)$, where A and B = average firing rate (Hz) during the self-administration session and the presession baseline, respectively.

If the behavioral interpretation were correct, one would expect that increments in reward magnitude, in the absence of drug, would have an effect on NAc firing that was similar to that of incrementing drug dose. We tested this prediction in one experiment. Three groups of animals were trained to self-administer 4%, 32%, or 64% sucrose (0.2 ml reinforcer) on an FR1 TO 60-sec schedule. Recordings were conducted on the 10th day of training. The number of neurons recorded at each of the three sucrose concentrations equaled 32, 51, and 38, respectively. Between-group comparisons showed that increasing sucrose concentration did not increase the prevalence or magnitude of session-long decreases that occurred during the sucrose self-administration session. However, there was some evidence that the prevalence of session-long increases were more prevalent at the 32% and 64% concentrations relative to the 4% concentration (Figure 2.22). The findings support the interpretation that the session-long decreases reflect something other than NAc tracking of reward magnitude. The relation between session-long increases and reward magnitude is less clear but may be positive.

Comparable findings to this dose-response study were found in a study of second-order self-administration. Four animals were trained on the FR10(FR10:S) procedure that we have already described. Animals earned a total of five cocaine reinforcers on this second-order schedule before initiating the second FR1 phase of the experiment. During the first second-order trial, the percent of neurons that showed tonic increases and decreases were similar and less than 30%. The percent of neurons that showed a significant decrease in firing relative to presession baseline increased across the subsequent four trials and increased further

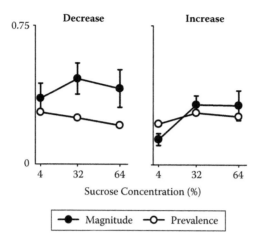

FIGURE 2.22 Tonic firing patterns during sucrose self-administration. Prevalence (left) and magnitude (right) of tonic firing patterns during FR1 self-administration of varying concentrations of sucrose. Data recorded during FR1 60-sec sessions.

during the second, FR1 TO 60-sec phase of the session. The progressive increase in inhibition occurred with progressive increases in drug level (Figure 2.23, top). The converse was true for increases in firing. Consistent with these changes, the average firing rate of neurons decreased across the five second-order trials (Figure 2.23, middle). Control analyses showed comparable results when between-trial comparisons of firing rate were limited to periods in which animals engaged in particular behaviors (Figure 2.24). The between-trial changes in firing thus did not depend on between-trial differences in prevalence of particular behaviors. The prevalence and magnitude of tonic increases and decreases in firing during the second phase of the session were comparable to those observed in previous experiments that used the FR1 TO 60-sec procedure. The study shows that inhibition of NAc neurons during drug-directed behavior is greater during periods of acute cocaine exposure than during drug-free periods. Moreover, inhibition of NAc neurons during drug-directed behavior increases as the level of cocaine rises (i.e., is dose dependent). These findings are consistent with those that one expects to observe if the acute pharmacological effect of cocaine on NAc neurons is inhibition.

A similar experiment was conducted in sucrose-trained animals. Four animals were trained to self-administer sucrose in the multiphase sessions. As for Aim 2 Exp 2, a second-order schedule FR10(FR10:S) was in effect during the first phase of the session and a first-order FR1 TO 60-sec schedule was in effect during the second phase of the session. During the first second-order trial, the prevalence of tonic increases and decreases were similar and less than 30%. Between the first second-order trial and the subsequent second-order trials, as well as the FR1 TO 60-sec phase, tonic changes in firing, and hence average firing rates, were highly stable (Figure 2.23 bottom). The findings are consistent with the interpretation that session-long NAc inhibition during sucrose-directed behavior is not affected by the presence

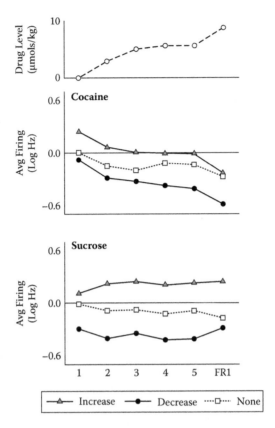

FIGURE 2.23 Average firing rate across 5 FR10(FR10:S) trials. Top and middle: Drug level and average firing rate of three groups of NAC neurons during 5 FR10:(FR10:S) cocaine-reinforced trials. The three groups correspond to neurons that showed either an increase, decrease, or no change in firing during the 1st FR10(FR10:S) trial relative to the presession baseline. Horizontal line represents presession baseline firing rate. Firing rates are normalized relative to baseline: $|A-B|/(A+B)$. Bottom: Comparable data are shown for a session in which animals self-administered sucrose in the same type of multiphase session.

versus absence of the primary reward. Inhibition of NAc firing during cocaine self-administration sessions is thus difficult to attribute to a normal nonpharmacological response of NAc neurons to reward exposure.

2.7.3 EVIDENCE THAT THE INHIBITORY EFFECTS OF SELF-ADMINISTERED COCAINE ARE ACTIVITY DEPENDENT

DA is hypothesized to have activity-dependent effects on the firing of NAc neurons. Specifically, it has been hypothesized that dopamine facilitates the transmission of strong excitatory input through the NAc, while simultaneously suppressing weak inputs (Rolls et al. 1984; Morgenson and Yim 1991; Pennartz et al. 1994; Pierce and Rebec 1995; Kiyatkin and Rebec 1996; Levine et al. 1996; O'Donnell and Grace

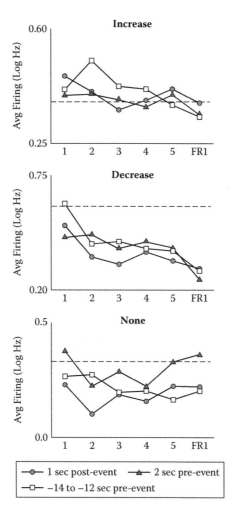

FIGURE 2.24 Firing during different periods of FR10(FR10:S) trials. Firing rates are shown for three groups of neurons: those that showed an increase (top), decrease (middle), or no change (bottom) in firing during the first FR10(FR10:S) trial. Firing rates were calculated during three periods of each FR10(FR10:S) trial: 2-sec prepress, when animals were pressing the response lever; the 1-sec postpress, when animals locomoted away from the lever, and the −14 to −12-sec prepress when animals engaged primarily in stereotypy. Legend refers to the press as the *event*. The horizontal line shows average firing during presession baseline. The figure shows that absolute firing rates varied across the behaviorally distinct periods; however, the type of change that each neuron group exhibited relative to the presession baseline was the same.

1996; Hernandez-Lopez et al. 1997). Given the importance of DA in cocaine effects on NAc neurons in anesthetized animal and slice studies, we have hypothesized that inhibition of NAc neurons during cocaine self-administration sessions might be activity dependent. Specifically, during cocaine self-administration sessions, neurons that receive excitatory afferent input associated with task-related events might be less susceptible to the inhibitory effects of self-administered cocaine. For example, neurons that are phasically activated during cocaine self-administration sessions might, as a group, show less inhibition than neurons that show no phasic excitatory activity. Evidence consistent with this expectation has been observed in a number of studies.

In one of the studies (Peoples et al. 2007b), rats were trained in a small number of short sessions to self-administer cocaine on an FR1 schedule of reinforcement (0.75 mg/kg/inf). Animals were then exposed to a 30-day regimen of daily long-access (LgA; 6 hr/day) cocaine self-administration. Recordings were conducted on the second, third, and 30th day of cocaine self-administration. Recorded neurons were sorted into three groups: (1) all those that showed a session-long increase in firing (session-activated neurons, e.g., Figure 2.13A); (2) those that showed a phasic excitatory response to a task event, including the cue that signals the onset of the session and the first and subsequent cocaine-reinforced lever presses (i.e., referred to as event-but-not-session-activated neurons; e.g., Figures 2.6, 2.15, 2.16); and (3) those that showed neither a session-long increase nor a response to any task event (i.e., task-nonactivated neurons).

Within each recording session, the three groups of neurons maintained different average firing rates (Figure 2.25, top versus middle versus bottom rows). Specifically, average firing rate of the session-activated neurons was significantly greater than average firing of the event-but-not-session-activated neurons, and the average firing rate of each of those neuron groups was greater than that of the task-nonactivated neurons. The different rates of average firing appeared to reflect a differential sensitivity of the neuron groups to acute inhibitory effects of cocaine. For example, the event-but-not-session-activated neurons, as a group, maintained

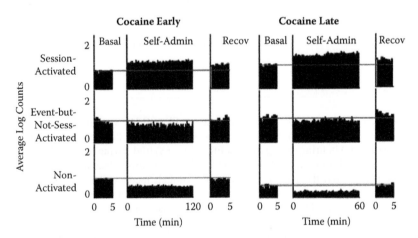

FIGURE 2.25 Average firing of groups of neurons during an early day (Day 2–3) and a late day (Day 30) of the 30-day regimen of long-access (6-hr) cocaine self-administration. Firing is shown during three periods for each session: (1) the last 5 min of the presession baseline period, (2) the last 1–2 hr of the self-administration session, and (3) the last 5 min of the postsession baseline period. Firing is shown for: (1) session-activated neurons (top), (2) event-but-not-session-activated neurons (middle), and (3) nonactivated neurons (bottom). Figure taken from Peoples et al., 2007, *Neuropsychopharmacology* 32:1141–1158.

average firing during the cocaine self-administration session relative to a presession baseline period. This was true even if phasic firing was excluded from calculations of firing rate. In contrast, the task-nonactivated neurons showed a significant decrease in average firing during the cocaine self-administration session relative to the pre-session baseline period. Similar findings were observed in an earlier study (Peoples et al. 2004). The findings are consistent with the interpretation that the acute inhibitory actions of self-administered cocaine are activity dependent, though additional studies are required to determine whether the activity-correlated changes in firing reflect activity-dependent, DA-mediated actions on the recorded neurons.

2.7.4 ACUTE PHARMACOLOGICAL EFFECTS OF SELF-ADMINISTERED COCAINE: SUMMARY, NOVEL FINDINGS, AND RESEARCH OPPORTUNITIES

Electrophysiological recordings generally show that experimenter-delivered cocaine inhibits spontaneous and glutamate-evoked firing of NAc neurons in both the slice and anesthetized animal preparations (Qiao et al. 1990; Uchimura and North 1990; White et al. 1993; Kiyatkin and Rebec 1996; Nicola et al. 1996). The findings of our behaving animal recordings are consistent with the interpretation that the primary pharmacological acute action of self-administered cocaine is also inhibition. However, the behaving animal studies show that not all neurons are inhibited during cocaine self-administration. In fact, a substantive portion of neurons show session-long increases in firing, phasic increases in firing, or both (Peoples et al. 1998b, 2007b). Moreover, the magnitude of inhibition induced by self-administered cocaine is reduced in neurons that receive excitatory afferent input (i.e., are phasically activated) relative to neurons that do not receive such afferent input during drug exposure. The substantial portion of noninhibited neurons and the possible activity-dependent nature of cocaine-induced inhibition are not predicted by the slice and anesthetized animal studies of acute cocaine effects on NAc neurons. The findings are likely to have important implications for understanding acute effects of cocaine that are relevant to cocaine addiction.

2.8 CHRONIC DRUG-INDUCED NEUROADAPTATIONS

2.8.1 SLICE AND ANESTHETIZED ANIMAL RECORDING STUDIES

Multiple electrophysiological studies have shown that a history of repeated cocaine exposure induces hypoactivity. In anesthetized animals, the responsiveness of NAc neurons to excitatory stimulation is reduced in animals with a history of experimenter-delivered cocaine (White 1992; White et al. 1995a). Similarly, in NAc slices, excitatory currents and whole cell excitability are decreased in rats with a history of cocaine exposure (Zhang et al. 1998). Moreover, repeated injections of cocaine can lead to a decrease in the ratio of AMPA to NMDA (i.e., AMPA:NMDA) currents, a mechanism linked to long-term-depression (LTD) (White et al. 1995b; Zhang et al. 1998; Thomas et al. 2001; Beurrier and Malenka 2002). Taken as a whole, these data are strongly consistent with the interpretation that a history of repeated cocaine is associated with decreases in excitatory synaptic and neuronal activity in the accumbens. If this were so, one would expect to observe comparable results in behaving animals with a history of cocaine self-administration. One would also expect that hypoactivity would correlate with increased propensity to seek and take drug.

The number of behaving animal recording studies that have investigated cocaine-induced plasticity is small. Nevertheless, the research that has been completed and that we will review exemplifies two approaches to testing for chronic effects of drug on neural activity in behaving animals including the following: (1) tracking the activity of individual neurons across days and (2) comparing average firing of groups of neurons recorded in animals exposed to different histories of drug self-administration. The research also exemplifies efforts to identify chronic drug effects that are related to the emergence of addiction-like patterns of behavior. Lastly, the work provides demonstrations of the methods used to differentiate chronic drug effects from normal plasticity associated with the nonpharmacological aspects of the experimental treatment.

2.8.2 RECORDINGS IN BEHAVING ANIMALS

In an early behaving-animal recording study of cocaine self-administration (Peoples et al. 1999a) we tested for evidence of hypoactivity in animals with a two-week history of cocaine self-administration. In that experiment, animals self-administered an intermediate (0.75 mg/kg/inf) dose of cocaine 6 hrs per day for 14–17 days. A recording session was conducted on days 2–3 and 14–17 of training. In this experiment we used a "within-neuron" approach to testing for changes in average firing of individual neurons. We identified neurons that were recorded on both recording days. These neurons were defined as those that met the following criteria. First, the population of waveforms recorded by a single microwire had to show evidence of a minimum inter-spike interval (ISI) consistent with the refractory period of a single neuron during both sessions. Second, the waveforms recorded by the same microwire had to be comparable between sessions (stability criterion). Specifically, for both days it was necessary for neural waveforms to have been defined by the same eight discrimination parameters and by a comparable range of variation in each of those eight parameters. To be included in the between-session comparisons, neural recordings had to be consistent with an additional criterion. Specifically, the neural waveforms had to be sufficient in amplitude on each day for the smallest discriminated waveforms to exceed the amplitude of the typical maximal fluctuations in the noiseband (completeness criterion). This requirement for a minimum waveform amplitude allowed us to verify that our ability to detect discharges did not change between days 2 and 15.

On day 2 of cocaine self-administration, 83 microwires recorded activity of single neurons. Of the 83 microwires, 53 recorded neural activity on both day 2 and day 15. Of those 53 wires, only 13 (24%) yielded neural records that met the criteria of stability and completeness. The majority of the 13 neurons showed a significant decrease in basal firing between the first and the second recording session. This finding is consistent with the interpretation that self-administered cocaine induces hypoactivity, as suggested by slice and anesthetized animal studies.

We characterized the hypoactivity further in another study, which employed a between-group approach. Rats self-administered cocaine in 30, daily, long-access (6-hr) sessions. Chronic extracellular recordings of the activity of individual NAc neurons were made during sessions 2–3 and 30 of the regimen (i.e., referred to as early versus late sessions). Each recording session included a presession baseline period, a cocaine self-administration session, and a postsession recovery period. Neurons recorded on each day were categorized into two groups, task-activated and task-nonactivated. Task-activated neurons were defined as all neurons that showed either of the following firing patterns: (1) a session-long increase in firing or (2) an excitatory change in firing time-locked to a discrete task-related event, including the cues that signaled the onset of the self-administration session and the operant behavior. All other neurons were defined as task-nonactivated neurons (Peoples et al. 2007b). About half of all recorded neurons in both the early and the late sessions were task-activated, while the other half were task-nonactivated. The average firing of the two groups of neurons was compared both within each of the self-administration sessions and across the two recording sessions.

The between-session comparisons showed that the presession average firing rate of the task-nonactivated group was significantly depressed on day 30 of self-administration relative to day 2–3. In contrast, the average presession firing rate of the task-activated group did not change between the early and the late self-administration sessions (Figure 2.26). The differential between-session change in firing was associated with the emergence of a significant difference between the baseline firing rates of the task-activated and task-nonactivated neurons and an increase in the ratio of firing rates of the activated neurons relative to the nonactivated neurons. Comparable findings were observed in similar between-session comparisons of the self-administration and postsession baseline recovery phases of the early and late recording sessions. Behavior and reward-related expectations differ among the presession baseline period, the self-administration session, and the postsession baseline period of the recording sessions. The similar differential between-session changes in firing among the different phases of the recording session showed that the differential changes in firing were not attributable to between-session changes in either behavior or specific expectations about the operant session (Figure 2.27). The differential changes in firing thus potentially reflect differential drug-induced neuroadaptations.

In an additional analysis of the same data, we more finely sorted neurons based on their response characteristics. Specifically, neurons in the activated group were subdivided into two groups: (1) neurons that showed a session-long increase (referred to as session-increase neurons), and (2) neurons that showed a phasic increase in relation to a task event but showed no session-long increase (referred to as task-but-not-session-activated). Replication of the original comparisons showed that three groups of neurons exhibited different between-session changes in firing. Moreover, the between-session changes in firing exhibited by the three subtypes of neurons were positively related to the change in firing exhibited by the neurons during the self-administration session relative to the drug-free baseline period (Figure 2.25). Session-increase neurons maintained elevated rates of firing throughout individual self-administration sessions relative to presession baseline and showed a trend to increase baseline firing rates between the early and the late sessions. Task-but-not-

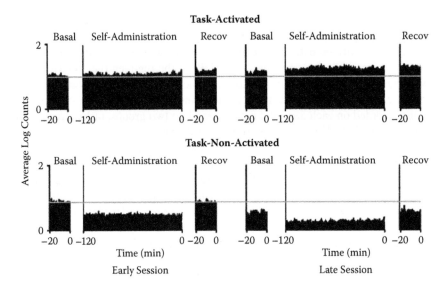

FIGURE 2.26 Population histograms showing the average baseline and self-administration firing rates exhibited by task-activated and task-nonactivated neurons during the early and the late sessions. Average firing of the task-activated neurons (top) and task-nonactivated neurons (bottom) during the early session (left) and the late session (right) are shown for the following periods: (1) last 20 min of the presession baseline period, (2) the last 2 hr of the self-administration session, and (3) the last 20 min of the postsession recovery periods. The solid gray horizontal lines across the histograms correspond to the average presession baseline firing rate during the early session and are shown to facilitate between-session comparisons of average firing rates. Figure taken from Peoples et al., 2007, *Neuropsychopharmacology* 32:1141–1158.

session-activated neurons maintained average firing rates during individual self-administration sessions that were similar to baseline firing rates and correspondingly showed no between-session change in baseline. Finally, task-nonactivated neurons were inhibited during self-administration sessions and showed a between-session decrease in basal activity. The findings are consistent with the interpretation that between-session changes in firing exhibited by NAc neurons are directionally consistent with the change in firing exhibited by the neurons during individual cocaine self-administration sessions, and thus with the hypothesis that neuroadaptations induced by self-administered cocaine develop in an activity-dependent manner.

In this study, two lines of evidence supported the interpretation that the differential changes in firing of the task-activated and task-nonactivated neurons were correlated with increases in the propensity of animals to seek and take drug. First, between the early and late self-administration sessions, there was a significant increase in the average rate of cocaine self-administration exhibited by all subjects (Figure 2.28). Second, during the late recording session, the average difference in basal firing rates between task-activated and task-nonactivated neurons was predictive of animals' drug-directed behavior during the later phases of the recording session. Animals with the greatest difference in basal firing between activated and nonactivated neurons

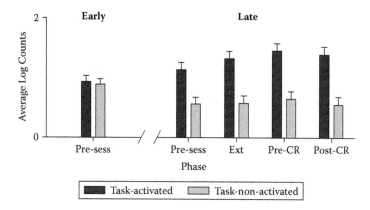

FIGURE 2.27 Average firing of task-activated and task-nonactivated neurons: no effect of the extinction and cue reinstatement procedure. Average firing of task-activated and task-non-activated neurons is plotted during the following periods: (1) 30 sec prior to presentation of the cue that signaled the onset of the early self-administration session (presession), (2) 30 sec before the start of the late session, (3) 29.5–30 min of the extinction procedure, (4) 30 sec prior to the start of the cue reinstatement period (pre-CR), and (5) 30 sec presentation of the drug-associated cues at the start of the cue reinstatement (post-CR) phase. Figure taken from Peoples et al., 2007, *Neuropsychopharmacology* 32:1141–1158.

consumed the greatest amount of cocaine during the self-administration session. The same animals also completed the highest number of lever-press responses during extinction and cue-reinstatement probes that were conducted later during the recording session (Figure 2.29). The finding is suggestive of a relationship between the difference in basal firing between task-activated and task-nonactivated neurons and the propensity of animals to engage in drug-directed behavior.

2.8.3 Effects of Extended Abstinence

In our previous study (Peoples et al. 2007b) we compared phasic firing time-locked to a number of cocaine task-related events between days 2–3 and 30 of LgA (long-access, 6-hr) cocaine self-administration. At the time of each recording, animals had been exposed to a short period of drug abstinence (≈18 hrs). The task-related firing patterns that were compared included firing patterns time-locked to a cue that signaled the onset of the session and phasic firing time-locked to the cocaine-reinforced lever press. We found no significant changes in either the prevalence or the magnitude of either type of event-related phasic firing. However, in two studies, Hollander and Carelli (2005, 2007) trained animals to self-administer cocaine in ShA (short-access, 2-hr) sessions and then compared task-related phasic firing patterns between animals exposed to either 1 or 30 days of abstinence. The compared firing patterns were similar to those evaluated in the Peoples et al. (2007b) study and included phasic firing time-locked to cocaine-associated cues presented to drug-free animals and operant responding during the cocaine self-administration session. Both measures of phasic activity were increased after 30 days of abstinence relative to 1 day of abstinence. Jones et al. (2008) showed that similar changes in phasic activity do not occur in

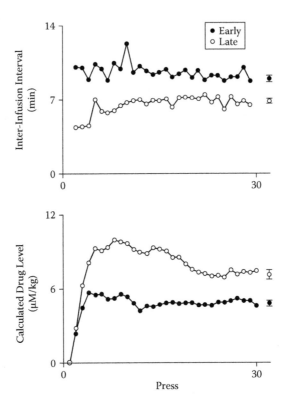

FIGURE 2.28   Average rate of drug intake for all the subjects during the early and the late sessions. At the top of the figure, average interinfusion interval (i.e., min between successive self-infusions) is plotted as a function of press number. At the bottom of the figure, average calculated drug level (µM/kg) at the time of the press is plotted as a function of lever-press. To the right of each plot is shown the average median value (± standard error) for the last 10 cocaine reinforced lever presses. Figure taken from Peoples et al., 2007, *Neuropsychopharmacology* 32:1141–1158.

animals trained to self-administer sucrose. The changes in firing observed in the Hollander and Carelli studies thus reflect cocaine-induced adaptations. The different findings in the Peoples et al. (2007b) experiment and the two Hollander and Carelli (2005, 2007) studies could reflect an effect of extended abstinence, which was present in the latter studies but not in the former study, though there were additional between-experiment differences (e.g., duration of the daily self-administration session).

2.8.4   CHRONIC DRUG-INDUCED NEUROADAPTATIONS: SUMMARY, NOVEL FINDINGS, AND RESEARCH OPPORTUNITIES

Electrophysiological recordings in slice and anesthetized animal preparations show that repeated exposure to cocaine leads to hypoactivity of NAc neurons (White 1992; White et al. 1993, 1995a; Zhang et al. 1998; Thomas et al. 2001; Beurrier and Malenka 2002). Consistent with this observation, behaving animal recordings

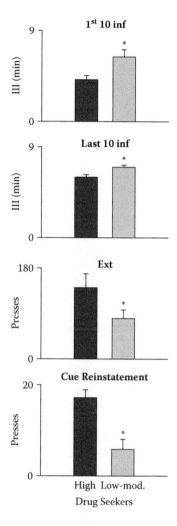

FIGURE 2.29   Various indices of drug seeking and taking during the late session are compared between the high and the low–moderate drug-seekers. The bar graphs show average measures (± standard error) of the drug seeking and taking behavior exhibited by animals that were identified as high drug-seekers and low–moderate drug-seekers. These measures include the following (from top to bottom of figure): (1) interinfusion interval (III) for the first 10 presses of the session, (2) interinfusion interval for presses 20–30, (3) total number of presses during the extinction phase, and (4) total number of presses during the cue reinstatement phase *$p<0.05$. Figure taken from Peoples et al., 2007, *Neuropsychopharmacology* 32:1141–1158.

show that repeated exposure to cocaine self-administration can lead to changes in basal firing that are consistent with the development of hypoactivity. However, those studies also show that the hypoactivity occurs only in neurons that do not exhibit excitatory changes in firing during cocaine self-administration sessions; moreover, excitatory changes in firing can be observed after extended periods of

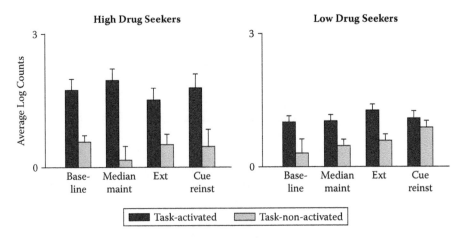

FIGURE 2.30 Average firing rates of task-activated and task-nonactivated neurons are compared between the high drug-seekers and the low-moderate drug-seekers. Average firing of task-activated and task-nonactivated neurons is plotted separately for animals that were defined as high (left panel) and low-moderate (right panel) drug-seekers. For each panel, average firing rates are shown for the following periods (left to right): (1) baseline (30-sec pre-S^D, i.e., 30 sec before presentation of the cue that signaled the onset of the session), (2) median prepress firing rate during the maintenance phase of the self-administration session, (3) min 29.5–30 of extinction, and (4) 30 sec post, after presentation of the drug-associated cue during the cue reinstatement phase. Figure taken from Peoples et al., 2007, *Neuropsychopharmacology* 32:1141–1158.

abstinence. The research also showed that the adaptations develop differentially among functionally distinct subtypes of neurons, perhaps because of the different activation patterns exhibited by those neurons during cocaine self-administration. Thus, recordings in behaving animals show that NAc firing patterns of behaving animals can undergo more complicated changes than was apparent based on the slice and anesthetized animal studies (though see Kourrich et al. 2008). The reviewed research emphasizes the additional leverage that the behaving-animal recording studies can add to electrophysiological investigations of addiction-relevant drug-induced adaptations. Indeed, characteristics of those adaptations may be subject to study only in behaving animals.

To date there are only four studies that have characterized the effects of repeated cocaine self-administration on single neuron activity in behaving animals. The behavioral procedures used in those studies were relatively simple. Nevertheless, the studies produced some findings that are relevant to current drug addiction hypotheses. In particular, the relative and absolute increments in NAc, phasic responses to drug-associated cues (Hollander and Carelli 2005, 2007; Peoples et al. 2007b) are in line with predictions of the incentive sensitization hypothesis and the differential neuroplasticity hypothesis. The research supports the idea that behaving animal recordings can contribute to testing drug addiction hypotheses. This is likely to be especially so, as the behavioral paradigms employed in the behaving animal recording studies become more sophisticated.

2.9 FUTURE DIRECTIONS

As in any discipline, the future of behaving animal recordings for investigating drug addiction is expected to hold new technical and procedural advances. Three broad categories of advance will be discussed in this section: advances in recording technology and data analysis, advances in behavioral assays, and the combination of chronic recording with other molecular techniques. This section overviews some but not all areas of ongoing and potential advances.

2.9.1 ADVANCES IN RECORDING TECHNIQUES AND ANALYSIS

Chronic implantation of arrays of electrodes was not widely used before the 1980s. Advances in computing technology enabled the use of multiple-electrode arrays and have governed many other advances in the technology since then. Multiple-electrode arrays have changed greatly since their first development. Most notably, the number of electrodes in these arrays has vastly increased and will continue to increase with advances in computing technology. In addition to the increasing number of electrodes, new configurations of electrodes allow for additional information to be obtained. Electrodes with multiple recording sites, such as stereotrodes and tetrodes, allow for the identification of single neurons on multiple channels, which can increase the confidence that neural spikes originate from a single neuron (Harris et al. 2000; Buzsaki 2004).

As technology has increased the number of electrodes that can be used concurrently to record neural activity, researchers are faced with the task of discriminating individual neuron signals from those channels. The process of discriminating recorded waveforms that belong to individual neurons is a time-consuming and somewhat subjective process (see Appendix I). Better spike-sorting algorithms are expected to speed the process up, make it more objective, and increase the number of successfully identified recorded neurons (Buzsaki 2004; Schmitzer-Torbert et al. 2005; Adamos et al. 2008; Chan et al. 2008).

Large numbers of simultaneously recorded neurons can reveal information that recordings of small numbers of neurons cannot. For example, neural activity that underlies a complex behavioral sequence may be observed as a population code across multiple neurons that respond at appropriate times in the behavioral sequence. Such a population code may only become apparent when looking at recordings of large numbers of neurons. Investigators are increasingly considering neurons as participants in networks of activity, and using analyses that observe the dynamics of large numbers of neurons simultaneously, rather than analyzing individual neurons independently (Petersen et al. 2002; Panzeri et al. 2003; Sanger 2003; Mazor and Laurent 2005; Averbeck et al. 2006; Fontanini and Katz 2006; Reddy and Kanwisher 2006; Lemon and Katz 2007).

2.9.2 ADVANCES IN BEHAVIORAL ASSAYS

Addiction is thought to include multiple stages, including controlled drug use, drug abuse, compulsive and uncontrolled drug seeking, abstinence, and relapse. An effort to extend the recording technique to self-administration paradigms that model these different phases will help to identify chronic drug-induced changes in neuronal activity that contribute to the progression of the disorder. The different findings of the Peoples et al. (2007) and Hollander and Carelli (2005, 2007) studies suggest that neurons in regions implicated in addiction, such as the NAc, may undergo a series of neuroadaptive changes in activity across phases of active drug use and extended abstinence. A recent nicotine self-administration study in our laboratory (K Guillem and LL Peoples, in press) supports this idea. Characterizing neural activity across phases of initial drug use, abstinence, and re-exposure could be very important to a full understanding of changes in neuronal activity that contribute to drug addiction (Guillem and Peoples, in press).

Greater application of the chronic recording procedure to studies of drug seeking would extend the utility of the method. Relapse prevention is a primary therapeutic endpoint in addiction treatment. As we have already described, there is evidence that drug-directed behavior under drug-free and drug-exposed conditions is mediated by somewhat different mechanisms. The multiphase seeking–taking procedure that we are developing and that was described in this chapter may be a useful paradigm for characterizing neural mechanisms associated with drug seeking under drug-free conditions. The method may be preferable relative to the common employed extinction–reinstatement procedures, which do not mimic conditions of abstinence and relapse in humans (i.e., humans do not undergo instrumental extinction). Application of the behaving-animal recording method to studies of chronic drug effects could be greatly advanced by these experimental design and procedural developments.

Progress in applying the chronic-behaving-animal recording procedure to investigating drug addiction hypotheses will depend to some degree on inclusion of more sophisticated behavioral approaches to the recording experiments. For example, procedures that temporally isolate events or motivational processes implicated by addiction hypotheses will be necessary if we are to identify the correlated neural responses and to test the effects of repeated drug exposure on those firing patterns. Studies of natural reward-directed behavior in animals with various histories of drug self-administration and abstinence will also be necessary additions to behaving animal studies of chronic drug effects that would be helpful in testing certain hypotheses. Finally, a variety of behavioral paradigms have been developed to test certain of the cognitive and motivational processes hypothesized to be affected by drug-induced adaptations and to be involved in addiction. Incorporating these paradigms in chronic animal recording studies would increase the power of those investigations to test addiction hypotheses.

2.9.3 GETTING AT MECHANISM: COMBINING CHRONIC RECORDING WITH OTHER TECHNIQUES

The identification of activity patterns of single neurons and groups of neurons that are related to drug seeking and drug reward is an important contribution of the chronic recording studies. However, the power of the research will be greatly enhanced if we can develop the methodology to more readily investigate the mechanisms that mediate those neural firing patterns. To do so it will be necessary to integrate the recording technique with additional neurochemical, neuropharmacological, biochemical, and molecular techniques.

Combining chronic recordings with molecular and cellular techniques, such as measurement of change in protein levels and receptor expression, will be useful for investigating the mechanisms of drug action. These studies most likely could best be conducted in separate sets of animals matched to the recording animals for drug exposure and so forth. The parallel studies could be useful in testing whether treatment regimens that engender certain changes in neural activity also engender particular biochemical and molecular adaptations that have been proposed to contribute to addiction.

Recent advances in molecular intervention methods using antisense and viruses now provide the unique opportunity to alter receptor functionality in a brain, region-specific manner. Application of these treatments at the time of surgery, could allow researchers to test specific hypotheses about receptor mechanisms and signal transduction pathways that might mediate acute and chronic effects of drug on neural activity.

Some of these suggested future directions are easier to implement than others. However, integration of the chronic recordings with the additional techniques will provide the opportunity to bridge the functional behaving-animal recording studies with other neuroscience fields of drug addiction. This could contribute significantly to developing a more complete picture of the mechanisms that mediate reward and addiction.

2.10   CONCLUSIONS

In the last 20–30 years, researchers have investigated the neurophysiological effects of acute and repeated exposure to self-administered drug. Much of the completed work has been conducted in slice and anesthetized animal preparations. Although these procedures can be used to conduct elegant mechanistic studies of acute and chronic drug effects on single neuron activity, drug actions and effects in these preparations do not always correspond with those that occur in behaving animals. This is because anesthesia and the absence of normal afferent input to neurons during the recording procedures can alter drug actions and effects. To address these issues it is necessary to characterize drug effects in behaving animals with a relevant history of drug self-administration.

The chronic-behaving-animal recording procedure can be used to characterize both acute and chronic effects of self-administered drug on individual neuron firing and to identify those particular drug-induced changes in neural activity that are associated with the emergence of addiction-like patterns of behavior. The method can also be used to test drug addiction hypotheses, which generally propose that drug-induced alterations in neural responses associated with specific motivational processes underlie the emergence of addiction-like behaviors. Though the chronic-behaving-animal recording method affords a number of novel research opportunities, the utility of the procedure in identifying pharmacological effects of addictive drugs depends on certain behavioral controls. Specifically, use of the chronic record-ing method to investigate acute pharmacological effects requires controls for afferent feedback associated with drug-induced changes in behavior and motivational pro-cesses. Similarly, use of the method in studies of chronic drug-induced adaptations requires controls for normal experiential plasticity that might occur with repeated exposure to the experimental procedures. Application of the technique to test addic-tion hypotheses will also depend on attention to various behavioral issues.

Thus far, the chronic-behaving-animal recording procedure has been applied pri-marily to studies of accumbal neural activity in animals trained to self-administer cocaine. A portion of those studies were reviewed in the present chapter. The research that was described exemplifies some of the methods that can be used to identify functional correlates of single neuron activity during drug seeking and taking, and to investigate acute and chronic effects of self-administered drug. The research that was described also highlighted a number of novel findings that have substantive implications for understanding mechanisms that mediate cocaine-directed behav-ior and cocaine addiction. Some of these include the following. First, task-related firing patterns of NAc neurons appear to be similar during self-administration of cocaine and natural, food, or fluid rewards. However, the neurons that exhibit the task-related firing patterns differ depending on the reward. Second, drug-associated cues and cocaine-directed behavior are associated with predominantly excitatory phasic (event-related) NAc activity under both drug-free and drug-exposed condi-tions. Both drug seeking and taking are also associated with sustained changes in average firing. These changes are almost exclusively excitatory in the absence of reinforcement. However, increases and decreases in firing occur when animals are exposed to conditioned and primary cocaine reinforcement, with decreases actually predominating during cocaine self-administration. Third, self-administered cocaine has primarily inhibitory acute effects on NAc activity, and repeated exposure to cocaine self-administration can decrease NAc basal activity. These findings are consistent with those of slice and anesthetized animal recording studies. However, the chronic recording studies showed that acute and chronic cocaine effects are more complex in behaving animals as compared with slice and anesthetized ani-mal preparations. Specifically, in behaving animals, neurons exhibit heterogeneous and activity-dependent responses to acute and chronic drug exposure. In fact, some neurons exhibit excitatory rather than inhibitory changes in firing. Finally, chronic recording studies have shown that NAc neurons exhibit changes in firing in response to repeated cocaine self-administration that are consistent with predictions of the incentive sensitization hypothesis and the differential neuroplasticity hypothesis. These findings demonstrate the utility of the chronic behaving animal recording method. Indeed there may be mechanisms that are critical to addiction which are only observable in the behaving animal studies.

Despite the advantages of the chronic-behaving-animal recording method, the procedure is currently not conducive to the same types of mechanistic studies that can be employed in acute and anesthetized electrophysiological studies of drug actions. Given the relative strengths and weaknesses of the different electrophysi-ological methods, complementary application of the techniques and consideration of findings from those studies as a whole will be important to understanding the drug effects that contribute to drug addiction. Slice and anesthetized animal methods can be used to identify drug effects and to characterize mechanisms of drug action. Recordings in behaving animals can be used to test for changes in neuronal activity that are predicted on the basis of the slice and anesthetized recordings, to identify additional drug effects that may be critical to addiction but absent in the typical slice and anesthetized animal studies, to test for changes in neural activity that are linked to particular addiction-like patterns of behavior, and to test certain predictions of drug addiction hypotheses. The advantages that can be gained by an integrative approach are demonstrated by the research described in the present chapter.

ACKNOWLEDGMENTS

LLP supported by NIH/NIDA P60 DA 005186 (CP O'Brien) and NIH/NIDA P50 DA 012756 (H Pettinati). KG supported by NIH Director's Bench-to-Bedside (LLP, Elliot Stein). AVK supported by DA-07241(CP O'Brien) and DA-021449 (AVK).

REFERENCES

Adamos DA, Kosmidis EK, Theophilidis G (2008) Performance evaluation of PCA-based spike sorting algorithms. *Comput Methods Programs Biomed* 91:232–244.

Ahmed SH, Koob GF (1998) Transition from moderate to excessive drug intake: change in hedonic set point. *Science* 282:298–300.

American Psychiatric Association (1994) *Diagnostic and statistical manual of mental disorders* (4th ed). Washington, D.C.: American Psychiatric Association.

Antkowiak B (1999) Different actions of general anesthetics on the firing patterns of neocortical neurons mediated by the GABA(A) receptor. *Anesthesiology* 91:500–511.

Antkowiak B, Helfrich-Forster C (1998) Effects of small concentrations of volatile anesthetics on action potential firing of neocortical neurons in vitro. Anesthesiology 88:1592–1605.

Apicella P, Ljungberg T, Scarnati E, Schultz W (1991) Responses to reward in monkey dorsal and ventral striatum. *Exp Brain Res* 85:491–500.

Arroyo M, Markou A, Robbins TW, Everitt BJ (1998) Acquisition, maintenance and reinstatement of intravenous cocaine self-administration under a second-order schedule of reinforcement in rats: effects of conditioned cues and continuous access to cocaine. Psychopharmacology (Berl) 140:331–344.

Averbeck BB, Latham PE, Pouget A (2006) Neural correlations, population coding and computation. *Nat Rev Neurosci* 7:358–366.

Ben-Shahar O, Keeley P, Cook M, Brake W, Joyce M, Nyffeler M, Heston R, Ettenberg A (2007) Changes in levels of D1, D2, or NMDA receptors during withdrawal from brief or extended daily access to IV cocaine. *Brain Res* 1131:220–228.

Beninger RJ (1983) The role of dopamine in locomotor activity and learning. *Brain Res* 287:173–196.

Berke JD, Hyman SE (2000) Addiction, dopamine, and the molecular mechanisms of memory. *Neuron* 25:515–532.

Beurrier C, Malenka RC (2002) Enhanced inhibition of synaptic transmission by dopamine in the nucleus accumbens during behavioral sensitization to cocaine. *J Neurosci* 22:5817–5822.

Bowman EM, Aigner TG, Richmond BJ (1996) Neural signals in the monkey ventral striatum related to motivation for juice and cocaine rewards. *J Neurophysiol* 75:1061–1073.

Breiter HC, Gollub RL, Weisskoff RM, Kennedy DN, Makris N, Berke JD, Goodman JM, et al. (1997) Acute effects of cocaine on human brain activity and emotion. *Neuron* 19:591–611.

Buzsaki G (2004) Large-scale recording of neuronal ensembles. *Nat Neurosci* 7:446–451.

Cador M, Robbins TW, Everitt BJ (1989) Involvement of the amygdala in stimulus-reward associations: interaction with the ventral striatum. *Neuroscience* 30:77–86.

Cador M, Taylor JR, Robbins TW (1991) Potentiation of the effects of reward-related stimuli by dopaminergic-dependent mechanisms in the nucleus accumbens. *Psychopharmacology* (Berl) 104:377–385.

Calu DJ, Roesch MR, Stalnaker TA, Schoenbaum G (2007) Associative encoding in posterior piriform cortex during odor discrimination and reversal learning. *Cereb Cortex* 17:1342–1349.

Cardinal RN, Parkinson JA, Lachenal G, Halkerston KM, Rudarakanchana N, Hall J, Morrison CH, Howes SR, Robbins TW, Everitt BJ (2002) Effects of selective excitotoxic lesions of the nucleus accumbens core, anterior cingulate cortex, and central nucleus of the amygdala on autoshaping performance in rats. *Behav Neurosci* 116:553–567.

Carelli RM (2000) Activation of accumbens cell firing by stimuli associated with cocaine delivery during self-administration. *Synapse* 35:238–242.

Carelli RM, Ijames SG (2000) Nucleus accumbens cell firing during maintenance, extinction, and reinstatement of cocaine self-administration behavior in rats. *Brain Res* 866:44–54.

Carelli RM, Ijames SG, Crumling AJ (2000) Evidence that separate neural circuits in the nucleus accumbens encode cocaine versus "natural" (water and food) reward. *J Neurosci* 20:4255–4266.

Carelli RM, King VC, Hampson RE, Deadwyler SA (1993) Firing patterns of nucleus accumbens neurons during cocaine self-administration in rats. *Brain Res* 626:14–22.

Carelli RM, Wightman RM (2004) Functional microcircuitry in the accumbens underlying drug addiction: insights from real-time signaling during behavior. *Curr Opin Neurobiol* 14:763–768.

Chan HL, Wu T, Lee ST, Fang SC, Chao PK, Lin MA (2008) Classification of neuronal spikes over the reconstructed phase space. *J Neurosci Methods* 168:203–211.

Chang JY, Janak PH, Woodward DJ (1998) Comparison of mesocorticolimbic neuronal responses during cocaine and heroin self-administration in freely moving rats. *J Neurosci* 18:3098–3115.

Chang JY, Sawyer SF, Lee RS, Woodward DJ (1994) Electrophysiological and pharmacological evidence for the role of the nucleus accumbens in cocaine self-administration in freely moving rats. *J Neurosci* 14:1224–1244.

Childress AR, Hole AV, Ehrman RN, Robbins SJ, McLellan AT, O'Brien CP (1993) Cue reactivity and cue reactivity interventions in drug dependence. *NIDA Res Monogr* 137:73–95.

Childress AR, Mozley PD, McElgin W, Fitzgerald J, Reivich M, O'Brien CP (1999) Limbic activation during cue-induced cocaine craving. *Am J Psychiatry* 156:11–18.

Colpaert FC (1987) Drug discrimination: methods of manipulation, measurement, and analysis. In: *Methods of assessing the reinforcing properties of abused drugs*, pp 341–372: Bozarth, M.A., Ed., New York: Springer.

Cromwell HC, Schultz W (2003) Effects of expectations for different reward magnitudes on neuronal activity in primate striatum. *J Neurophysiol* 89:2823–2838.

Deadwyler SA (1986) Electrophysiological investigations of drug influences in the behaving animal. In: *Modern methods in pharmacology. Vol. 3: Electrophysiological techniques in pharmacology*: Geller, H.M., Ed., Chap 1.

Deroche-Gamonet V, Belin D, Piazza PV (2004) Evidence for addiction-like behavior in the rat. *Science* 305:1014–1017.

Di Chiara G, Acquas E, Carboni E (1992) Drug motivation and abuse: a neurobiological perspective. *Ann N Y Acad Sci* 654:207–219.

Dickinson A, Wood N, Smith JW (2002) Alcohol seeking by rats: action or habit? *Q J Exp Psychol B* 55:331–348.

Everitt BJ, Cador M, Robbins TW (1989) Interactions between the amygdala and ventral striatum in stimulus-reward associations: studies using a second-order schedule of sexual reinforcement. *Neuroscience* 30:63–75.

Everitt BJ, Dickinson A, Robbins TW (2001) The neuropsychological basis of addictive behaviour. *Brain Res Brain Res Rev* 36:129–138.

Everitt BJ, Wolf ME (2002) Psychomotor stimulant addiction: a neural systems perspective. *J Neurosci* 22:3312–3320.

Fink-Jensen A, Ingwersen SH, Nielsen PG, Hansen L, Nielsen EB, Hansen AJ (1994) Halothane anesthesia enhances the effect of dopamine uptake inhibition on interstitial levels of striatal dopamine. *Naunyn Schmiedebergs Arch Pharmacol* 350:239–244.

Floresco SB, West AR, Ash B, Moore H, Grace AA (2003) Afferent modulation of dopamine neuron firing differentially regulates tonic and phasic dopamine transmission. *Nat Neurosci* 6:968–973.

Fontanini A, Katz DB (2006) State-dependent modulation of time-varying gustatory responses. *J Neurophysiol* 96:3183–3193.

Fuchs RA, Branham RK, See RE (2006) Different neural substrates mediate cocaine seeking after abstinence versus extinction training: a critical role for the dorsolateral caudate-putamen. *J Neurosci* 26:3584–3588.

Garavan H, Pankiewicz J, Bloom A, Cho JK, Sperry L, Ross TJ, Salmeron BJ, Risinger R, Kelley D, Stein EA (2000) Cue-induced cocaine craving: neuroanatomical specificity for drug users and drug stimuli. *Am J Psychiatry* 157:1789–1798.

Gawin FH, Kleber HD (1986) Abstinence symptomatology and psychiatric diagnosis in cocaine abusers. Clinical observations. *Arch Gen Psychiatry* 43:107–113.

Ghitza UE, Fabbricatore AT, Prokopenko V, Pawlak AP, West MO (2003) Persistent cue-evoked activity of accumbens neurons after prolonged abstinence from self-administered cocaine. *J Neurosci* 23:7239–7245.

Grant S, London ED, Newlin DB, Villemagne VL, Liu X, Contoreggi C, Phillips RL, Kimes AS, Margolin A (1996) Activation of memory circuits during cue-elicited cocaine craving. *Proc Natl Acad Sci U S A* 93:12040–12045.

Green JD (1958) A simple microelectrode for recording from the central nervous system. *Nature* 182:962.

Griffiths RR, Bigelow GE, Henningfield JE (1980) Animal and human drug-taking behavior. In: *Advances in Substance Abuse Behavioral and Biological Research*, Chap. 3: Mello NK, Ed., JAI Press: Greenwich, CT.

Grimm JW, See RE (2000) Dissociation of primary and secondary reward-relevant limbic nuclei in an animal model of relapse. *Neuropsychopharmacology* 22:473–479.

Gullienk K, Peoples LL. Progressive and lasting amplification of accumbal nicotine-seeking neural signals. *J Neurosci*.

Harris KD, Henze DA, Csicsvari J, Hirase H, Buzsaki G (2000) Accuracy of tetrode spike separation as determined by simultaneous intracellular and extracellular measurements. *J Neurophysiol* 84:401–414.

Hassani OK, Cromwell HC, Schultz W (2001) Influence of expectation of different rewards on behavior-related neuronal activity in the striatum. *J Neurophysiol* 85:2477–2489.

Hemby SE, Co C, Koves TR, Smith JE, Dworkin SI (1997) Differences in extracellular dopamine concentrations in the nucleus accumbens during response-dependent and response-independent cocaine administration in the rat. *Psychopharmacology* (Berl) 133:7–16.

Henry DJ, White FJ (1995) The persistence of behavioral sensitization to cocaine parallels enhanced inhibition of nucleus accumbens neurons. *J Neurosci* 15:6287–6299.

Hernandez-Lopez S, Bargas J, Surmeier DJ, Reyes A, Galarraga E (1997) D1 receptor activation enhances evoked discharge in neostriatal medium spiny neurons by modulating an L-type Ca2+ conductance. *J Neurosci* 17:3334–3342.

Hollander JA, Carelli RM (2005) Abstinence from cocaine self-administration heightens neural encoding of goal-directed behaviors in the accumbens. *Neuropsychopharmacology* 30:1464–1474.

Hollander JA, Carelli RM (2007) Cocaine-associated stimuli increase cocaine seeking and activate accumbens core neurons after abstinence. *J Neurosci* 27:3535–3539.

Hollerman JR, Tremblay L, Schultz W (1998) Influence of reward expectation on behavior-related neuronal activity in primate striatum. *J Neurophysiol* 80:947–963.

Hope BT, Crombag HS, Jedynak JP, Wise RA (2005) Neuroadaptations of total levels of adenylate cyclase, protein kinase A, tyrosine hydroxylase, cdk5 and neurofilaments in the nucleus accumbens and ventral tegmental area do not correlate with expression of sensitized or tolerant locomotor responses to cocaine. *J Neurochem* 92:536–545.

Hu XT, Basu S, White FJ (2004) Repeated cocaine administration suppresses HVA-Ca2+ potentials and enhances activity of K+ channels in rat nucleus accumbens neurons. *J Neurophysiol* 92:1597–1607.

Hu XT, Ford K, White FJ (2005) Repeated cocaine administration decreases calcineurin (PP2B) but enhances DARPP-32 modulation of sodium currents in rat nucleus accumbens neurons. *Neuropsychopharmacology* 30:916–926.

Hu XT, White FJ (1994) Loss of D1/D2 dopamine receptor synergisms following repeated administration of D1 or D2 receptor selective antagonists: electrophysiological and behavioral studies. *Synapse* 17:43–61.

Hurd YL, Kehr J, Ungerstedt U (1988) In vivo microdialysis as a technique to monitor drug transport: correlation of extracellular cocaine levels and dopamine overflow in the rat brain. *J Neurochem* 51:1314–1316.

Ito R, Dalley JW, Robbins TW, Everitt BJ (2002) Dopamine release in the dorsal striatum during cocaine-seeking behavior under the control of a drug-associated cue. *J Neurosci* 22:6247–6253.

Ito R, Dalley JW, Howes SR, Robbins TW, Everitt BJ (2000) Dissociation in conditioned dopamine release in the nucleus accumbens core and shell in response to cocaine cues and during cocaine-seeking behavior in rats. *J Neurosci* 20:7489–7495.

Iversen SD, Koob GF (1977) Behavioral implications of dopaminergic neurons in the mesolimbic system. *Adv Biochem Psychopharmacol* 16:209–214.

Janak PH (2002) Multichannel neural ensemble recording during alcohol self-administration. In: *Methods for alcohol-related neuroscience research*, pp 243–259: Liu, Y. and Lovinger, D.M., Eds., CRC Press: Boca Raton, FL.

Janak PH (2003) Application of many-neuron microelectrode array recording and the study of reward seeking behavior. In: *Methods in drug abuse research: cellular and circuit level analyses,* Waterhouse BD, Ed., CRC Press: Boca Raton, FL.

Johanson CE, Balster RL (1978) A summary of the results of a drug self-administration study using substitution procedures in rhesus monkeys. *Bull Narc* 30:43–54.

Johanson CE, Schuster CR (1975) A choice procedure for drug reinforcers: cocaine and methylphenidate in the rhesus monkey. *J Pharmacol Exp Ther* 193:676–688.

Jones JL, Wheeler RA, Carelli RM (2008) Behavioral responding and nucleus accumbens cell firing are unaltered following periods of abstinence from sucrose. *Synapse* 62:219–228.

Kalivas PW (2004) Recent understanding in the mechanisms of addiction. *Curr Psychiatry Rep* 6:347–351.

Kalivas PW, O'Brien C (2008) Drug addiction as a pathology of staged neuroplasticity. *Neuropsychopharmacology* 33:166–180.

Kelley AE, Stinus L (1985) Disappearance of hoarding behavior after 6-hydroxydopamine lesions of the mesolimbic dopamine neurons and its reinstatement with L-dopa. *Behav Neurosci* 99:531–545.

Kiyatkin EA, Rebec GV (1996) Dopaminergic modulation of glutamate-induced excitations of neurons in the neostriatum and nucleus accumbens of awake, unrestrained rats. *J Neurophysiol* 75:142–153.

Koob GF, Le Moal M (1997) Drug abuse: hedonic homeostatic dysregulation. *Science* 278:52–58.

Koob GF, Le Moal M (2001) Drug addiction, dysregulation of reward, and allostasis. *Neuropsychopharmacology* 24:97–129.

Koob GF, Le Moal M (2008) Addiction and the brain antireward system. *Annu Rev Psychol* 59:29–53.

Koob GF, Nestler EJ (1997) The neurobiology of drug addiction. *J Neuropsychiatry Clin Neurosci* 9:482–497.

Koob GF, Stinus L, Le Moal M, Bloom FE (1989) Opponent process theory of motivation: neurobiological evidence from studies of opiate dependence. *Neurosci Biobehav Rev* 13:135–140.

Kourrich S, Rothwell PE, Klug JR, and Thomas MJ (2007) Cocaine experience controls bidirectional synaptic plasticity in the nucleus accumbens. *J Neurosci* 27, 7921-7928.

Kravitz AV, Moorman DE, Simpson A, Peoples LL (2006) Session-long modulations of accumbal firing during sucrose-reinforced operant behavior. *Synapse* 60:420–428.

Kravitz AV, Peoples LL (2008) Background firing rates of orbitofrontal neurons reflect specific characteristics of operant sessions and modulate phasic responses to reward-associated cues and behavior. *J Neurosci* 28:1009–1018.

Lavoie AM, Mizumori SJ (1994) Spatial, movement- and reward-sensitive discharge by medial ventral striatum neurons of rats. *Brain Res* 638:157–168.

Lemon CH, Katz DB (2007) The neural processing of taste. *BMC Neurosci* 8 Suppl 3:S5.

Levine MS, Li Z, Cepeda C, Cromwell HC, Altemus KL (1996) Neuromodulatory actions of dopamine on synaptically-evoked neostriatal responses in slices. *Synapse* 24:65–78.

Lu L, Dempsey J, Shaham Y, Hope BT (2005) Differential long-term neuroadaptations of glutamate receptors in the basolateral and central amygdala after withdrawal from cocaine self-administration in rats. *J Neurochem* 94:161–168.

Mantsch JR, Ho A, Schlussman SD, Kreek MJ (2001) Predictable individual differences in the initiation of cocaine self-administration by rats under extended-access conditions are dose-dependent. *Psychopharmacology* (Berl) 157:31–39.

Markou A, Arroyo M, Everitt BJ (1999) Effects of contingent and non-contingent cocaine on drug-seeking behavior measured using a second-order schedule of cocaine reinforcement in rats. *Neuropsychopharmacology* 20:542–555.

Mazor O, Laurent G (2005) Transient dynamics versus fixed points in odor representations by locust antennal lobe projection neurons. *Neuron* 48:661–673.

Mereu G, Casu M, Gessa GL (1983) (-)-Sulpiride activates the firing rate and tyrosine hydroxylase activity of dopaminergic neurons in unanesthetized rats. *Brain Res* 264:105–110.

Misra AL, Pontani RB, Mule SJ (1976) [3H]-Noncocaine and [3H]-pseudococaine: effect of N-demethylation and C2-epimerization of cocaine on its pharmacokinetics in the rat. *Experientia* 32:895–897.

Morgenson G, Yim C (1991) Neuromodulatory functions of the mesolimbic dopamine system: electrophysiological and behavioral studies. In: *The mesolimbic dopamine system: from motivation to action*, Willner P, Scheel-Kruger J, Eds., New York: Wiley.

Moxon KA (1999) Multichannel electrode design: considerations for different application. In: *Methods for Neural Ensemble Recordings*, Nicolelis, M.A.L., Ed., Boca Raton, FL: CRC Press.

Nasif FJ, Hu XT, White FJ (2005) Repeated cocaine administration increases voltage-sensitive calcium currents in response to membrane depolarization in medial prefrontal cortex pyramidal neurons. *J Neurosci* 25:3674–3679.

Nestler EJ (2001) Molecular basis of long-term plasticity underlying addiction. *Nat Rev Neurosci* 2:119–128.

Nicola SM, Kombian SB, Malenka RC (1996) Psychostimulants depress excitatory synaptic transmission in the nucleus accumbens via presynaptic D1-like dopamine receptors. *J Neurosci* 16:1591–1604.

Nicola SM, Surmeier J, Malenka RC (2000) Dopaminergic modulation of neuronal excitability in the striatum and nucleus accumbens. *Annu Rev Neurosci* 23:185–215.

Nicola SM, Yun IA, Wakabayashi KT, Fields HL (2004) Firing of nucleus accumbens neurons during the consummatory phase of a discriminative stimulus task depends on previous reward predictive cues. *J Neurophysiol* 91:1866–1882.

O'Brien CP (2005) Anticraving medications for relapse prevention: a possible new class of psychoactive medications. *Am J Psychiatry* 162:1423–1431.

O'Brien CP, Childress AR, McLellan T, Ehrman R (1990) Integrating systemic cue exposure with standard treatment in recovering drug dependent patients. *Addict Behav* 15:355–365.

O'Brien CP, Childress AR, Ehrman R, Robbins SJ (1998) Conditioning factors in drug abuse: can they explain compulsion? *J Psychopharmacol* 12:15–22.

O'Donnell P (2003) Dopamine gating of forebrain neural ensembles. *Eur J Neurosci* 17:429–435.

O'Donnell P, Grace AA (1996) Dopaminergic reduction of excitability in nucleus accumbens neurons recorded in vitro. *Neuropsychopharmacology* 15:87–97.

O'Donnell P, Greene J, Pabello N, Lewis BL, Grace AA (1999) Modulation of cell firing in the nucleus accumbens. *Ann N Y Acad Sci* 877:157–175.

Olmstead MC, Lafond MV, Everitt BJ, Dickinson A (2001) Cocaine seeking by rats is a goal-directed action. *Behav Neurosci* 115:394–402.

Olmstead MC, Parkinson JA, Miles FJ, Everitt BJ, Dickinson A (2000) Cocaine-seeking by rats: regulation, reinforcement and activation. *Psychopharmacology* (Berl) 152:123–131.

Overton DA (1987) Applications and limitations of the drug discrimination method for the study of drug abuse. In: *Methods of assession the reinforcing properties of abused drugs*, pp 291–340: Bozarth, M.A., Ed., New York: Springer.

Panlilio LV, Goldberg SR (2007) Self-administration of drugs in animals and humans as a model and an investigative tool. *Addiction* 102:1863–1870.

Panzeri S, Pola G, Petersen RS (2003) Coding of sensory signals by neuronal populations: the role of correlated activity. *Neuroscientist* 9:175–180.

Paxinos G, Watson C (2004) *The rat brain in stereotaxic coordinates.* 4th Ed. New York: Elsevier.

Peltier RL, Guerin GF, Dorairaj N, Goeders NE (2001) Effects of saline substitution on responding and plasma corticosterone in rats trained to self-administer different doses of cocaine. *J Pharmacol Exp Ther* 299:114–120.

Pennartz CM, Groenewegen HJ, Lopes da Silva FH (1994) The nucleus accumbens as a complex of functionally distinct neuronal ensembles: an integration of behavioural, electrophysiological and anatomical data. *Prog Neurobiol* 42:719–761.

Peoples LL (2003) Application of chronic extracellular recording to studies of drug self-administration. In: *Methods in drug abuse research: cellular and circuits level analyses*, pp 161–211: Waterhouse, B.D., Ed., Boca Raton, FL: CRC Press.

Peoples LL, Bibi R, West MO (1994) Effects of intravenous selfadministered cocaine on single cell activity in the nucleus accumbens of the rat. In: Problems of drug dependence, 1993: proceedings of the 55th annual scientific meeting. The College on Problems of Drug

Dependence, Inc. Vol II, p pp 326. Washington, D.C.: Superintendent of Documents, United States Government Printing Office. National Institute on Drug Abuse Research Monograph 141.: Harris L, ed.

Peoples LL, Cavanaugh D (2003) Differential changes in signal and background firing of accumbal neurons during cocaine self-administration. *J Neurophysiol* 90:993–1010.

Peoples LL, Gee F, Bibi R, West MO (1998b) Phasic firing time locked to cocaine self-infusion and locomotion: dissociable firing patterns of single nucleus accumbens neurons in the rat. *J Neurosci* 18:7588–7598.

Peoples LL, Kravitz AV, Guillem K (2007a) The role of accumbal hypoactivity in cocaine addiction. *ScientificWorldJournal* 7:22–45.

Peoples LL, Kravitz AV, Lynch KG, Cavanaugh DJ (2007b) Accumbal neurons that are activated during cocaine self-administration are spared from inhibitory effects of repeated cocaine self-administration. *Neuropsychopharmacology* 32:1141–1158.

Peoples LL, Lynch KG, Lesnock J, Gangadhar N (2004) Accumbal neural responses during the initiation and maintenance of intravenous cocaine self-administration. *J Neurophysiol* 91:314–323.

Peoples LL, Uzwiak AJ, Gee F, Fabbricatore AT, Muccino KJ, Mohta BD, West MO (1999b) Phasic accumbal firing may contribute to the regulation of drug taking during intravenous cocaine self-administration sessions. *Ann N Y Acad Sci* 877:781–787.

Peoples LL, Uzwiak AJ, Gee F, West MO (1997) Operant behavior during sessions of intravenous cocaine infusion is necessary and sufficient for phasic firing of single nucleus accumbens neurons. *Brain Res* 757:280–284.

Peoples LL, Uzwiak AJ, Gee F, West MO (1999a) Tonic firing of rat nucleus accumbens neurons: changes during the first 2 weeks of daily cocaine self-administration sessions. *Brain Res* 822:231–236.

Peoples LL, Uzwiak AJ, Guyette FX, West MO (1998a) Tonic inhibition of single nucleus accumbens neurons in the rat: a predominant but not exclusive firing pattern induced by cocaine self-administration sessions. *Neuroscience* 86:13–22.

Peoples LL, West MO (1996) Phasic firing of single neurons in the rat nucleus accumbens correlated with the timing of intravenous cocaine self-administration. *J Neurosci* 16:3459–3473.

Petersen RS, Panzeri S, Diamond ME (2002) Population coding in somatosensory cortex. *Curr Opin Neurobiol* 12:441–447.

Pettit HO and Justice JB Jr. (1989) Dopamine in the nucleus accumbens during cocaine self-administration as studied by in vivo microdialysis. *Pharmacol Biochem Behav* 34: 899–904.

Pickens R, Thompson T (1968) Cocaine-reinforced behavior in rats: effects of reinforcement magnitude and fixed-ratio size. *J Pharmacol Exp Ther* 161:122–129.

Pierce RC, Rebec GV (1995) Iontophoresis in the neostriatum of awake, unrestrained rats: differential effects of dopamine, glutamate and ascorbate on motor- and nonmotor-related neurons. *Neuroscience* 67:313–324.

Qiao JT, Dougherty PM, Wiggins RC, Dafny N (1990) Effects of microiontophoretic application of cocaine, alone and with receptor antagonists, upon the neurons of the medial prefrontal cortex, nucleus accumbens and caudate nucleus of rats. *Neuropharmacology* 29:379–385.

Ranck JB, Jr., Kubie JL, Fox SE, Wolfson S, Muller RU (1983) Single neuron recording in behaving mammal: bridging the gap between neuronal events and sensor-behavioral variables. In: *Behavioral approches to brain research*, Chap. 5, Robinson, T.E., Ed., New York: Oxford University Press.

Reddy L, Kanwisher N (2006) Coding of visual objects in the ventral stream. *Curr Opin Neurobiol* 16:408–414.

Richardson NR, Roberts DC (1996) Progressive ratio schedules in drug self-administration studies in rats: a method to evaluate reinforcing efficacy. *J Neurosci Methods* 66:1–11.

Robbins TW (1978) The acquisition of responding with conditioned reinforcement: effects of pipradrol, methylphenidate, d-amphetamine, and nomifensine. *Psychopharmacology* (Berl) 58:79–87.

Robbins TW, Everitt BJ (1982) Functional studies of the central catecholamines. *Int Rev Neurobiol* 23:303–365.

Robinson TE, Berridge KC (1993) The neural basis of drug craving: an incentive-sensitization theory of addiction. *Brain Res Brain Res Rev* 18:247–291.

Robinson TE, Berridge KC (2003) Addiction. *Annu Rev Psychol* 54:25–53.

Roesch MR, Stalnaker TA, Schoenbaum G (2007) Associative encoding in anterior piriform cortex versus orbitofrontal cortex during odor discrimination and reversal learning. *Cereb Cortex* 17:643–652.

Rolls ET, Thorpe SJ, Boytim M, Szabo I, Perrett DI (1984) Responses of striatal neurons in the behaving monkey. 3. Effects of iontophoretically applied dopamine on normal responsiveness. *Neuroscience* 12:1201–1212.

Salomone J (1992) Report on Australian national conference: "Surrogacy—in whose interest?" Melbourne, February 1991. *Issues Reprod Genet Eng* 5:79–94.

Sanger TD (2003) Neural population codes. *Curr Opin Neurobiol* 13:238–249.

Schmitzer-Torbert N, Jackson J, Henze D, Harris K, Redish AD (2005) Quantitative measures of cluster quality for use in extracellular recordings. *Neuroscience* 131:1–11.

Schultz W (2000) Multiple reward signals in the brain. *Nat Rev Neurosci* 1:199–207.

Schultz W, Apicella P, Ljungberg T (1993) Responses of monkey dopamine neurons to reward and conditioned stimuli during successive steps of learning a delayed response task. *J Neurosci* 13:900–913.

Schultz W, Apicella P, Scarnati E, Ljungberg T (1992) Neuronal activity in monkey ventral striatum related to the expectation of reward. *J Neurosci* 12:4595–4610.

Self DW, Choi KH, Simmons D, Walker JR, Smagula CS (2004) Extinction training regulates neuroadaptive responses to withdrawal from chronic cocaine self-administration. *Learn Mem* 11:648–657.

Shaham Y, Shalev U, Lu L, De Wit H, Stewart J (2003) The reinstatement model of drug relapse: history, methodology and major findings. *Psychopharmacology* (Berl) 168:3–20.

Shalev U, Grimm JW, Shaham Y (2002) Neurobiology of relapse to heroin and cocaine seeking: a review. *Pharmacol Rev* 54:1–42.

Shidara M, Aigner TG, Richmond BJ (1998) Neuronal signals in the monkey ventral striatum related to progress through a predictable series of trials. *J Neurosci* 18:2613–2625.

Smith JE, Co C, Freeman ME, Lane JD (1982) Brain neurotransmitter turnover correlated with morphine-seeking behavior of rats. *Pharmacol Biochem Behav* 16:509–519.

Smith JE, Co C, Freeman ME, Sands MP, Lane JD (1980) Neurotransmitter turnover in rat striatum is correlated with morphine self-administration. *Nature* 287:152–154.

Spear DJ, Katz JL (1991) Cocaine and food as reinforcers: effects of reinforcer magnitude and response requirement under second-order fixed-ratio and progressive-ratio schedules. *J Exp Anal Behav* 56:261–275.

Stalnaker TA, Roesch MR, Franz TM, Calu DJ, Singh T, Schoenbaum G (2007) Cocaine-induced decision-making deficits are mediated by miscoding in basolateral amygdala. *Nat Neurosci* 10:949–951.

Stewart J, deWit H (1987) Reinstatement of drug-taking behavior as a method of assessing incentive motivational properties of drugs. In: *Methods of assessing the reinforcing properties of abused drugs*, pp 211–228: Bozarth, M.A., Ed., New York: Springer.

Stewart J, de Wit H, Eikelboom R (1984) Role of unconditioned and conditioned drug effects in the self-administration of opiates and stimulants. *Psychol Rev* 91:251–268.

Taha SA, Nicola SM, Fields HL (2007) Cue–evoked encoding of movement planning and execution in the rat nucleus accumbens. *J Physiol* 584:801-818.

Thomas MJ, Beurrier C, Bonci A, Malenka RC (2001) Long-term depression in the nucleus accumbens: a neural correlate of behavioral sensitization to cocaine. *Nat Neurosci* 4:1217–1223.

Thomsen M, Caine SB (2006) Cocaine self-administration under fixed and progressive ratio schedules of reinforcement: comparison of C57BL/6J, 129X1/SvJ, and 129S6/SvEvTac inbred mice. *Psychopharmacology* (Berl) 184:145–154.

Thomas MJ, Kalivas PW, Shaham Y (2008) Neuroplasticity in the mesolimbic dopamine system and cocaine addiction. *Br J Pharmacol* 154(2):327–342.

Tiffany ST (1999) Cognitive concepts of craving. *Alcohol Res Health* 23:215–224.

Uchimura N, North RA (1990) Actions of cocaine on rat nucleus accumbens neurones in vitro. *Br J Pharmacol* 99:736–740.

Uzwiak AJ, Guyette FX, West MO, Peoples LL (1997) Neurons in accumbens subterritories of the rat: phasic firing time-locked within seconds of intravenous cocaine self-infusion. *Brain Res* 767:363–369.

Vanderschuren LJ, Everitt BJ (2004) Drug seeking becomes compulsive after prolonged cocaine self-administration. *Science* 305:1017–1019.

Volkow ND, Wang GJ, Fowler JS, Hitzemann R, Angrist B, Gatley SJ, Logan J, Ding YS, Pappas N (1999) Association of methylphenidate-induced craving with changes in right striato-orbitofrontal metabolism in cocaine abusers: implications in addiction. *Am J Psychiatry* 156:19–26.

Wan X, Peoples LL (2008) Amphetamine exposure enhances accumbal responses to reward-predictive stimuli in a pavlovian conditioned approach task. *J Neurosci* 28(30):7501–7512.

Weddington WW, Brown BS, Haertzen CA, Cone EJ, Dax EM, Herning RI, Michaelson BS (1990) Changes in mood, craving, and sleep during short-term abstinence reported by male cocaine addicts. A controlled, residential study. *Arch Gen Psychiatry* 47:861–868.

West AR, Floresco SB, Charara A, Rosenkranz JA, Grace AA (2003) Electrophysiological interactions between striatal glutamatergic and dopaminergic systems. *Ann N Y Acad Sci* 1003:53–74.

West MO, Peoples LL, Michael AJ, Chapin JK, Woodward DJ (1997) Low-dose amphetamine elevates movement-related firing of rat striatal neurons. *Brain Res* 745:331–335.

White FJ, Henry, D.J, Jeziorski, M., and Ackerman, J.M. (1992) *Electrophysiologic effects of cocaine within the mesoaccumbens and mesocortical dopamine systems.* Boca Raton, FL: CRC Press.

White FJ, Hu XT, Henry DJ (1993) Electrophysiological effects of cocaine in the rat nucleus accumbens: microiontophoretic studies. *J Pharmacol Exp Ther* 266:1075–1084.

White FJ, Hu XT, Zhang XF (1998) Neuroadaptations in nucleus accumbens neurons resulting from repeated cocaine administration. *Adv Pharmacol* 42:1006–1009.

White FJ, Hu XT, Zhang XF, Wolf ME (1995a) Repeated administration of cocaine or amphetamine alters neuronal responses to glutamate in the mesoaccumbens dopamine system. *J Pharmacol Exp Ther* 273:445–454.

White SR, Harris GC, Imel KM, Wheaton MJ (1995b) Inhibitory effects of dopamine and methylenedioxymethamphetamine (MDMA) on glutamate-evoked firing of nucleus accumbens and caudate/putamen cells are enhanced following cocaine self-administration. *Brain Res* 681:167–176.

White FJ, Kalivas PW (1998) Neuroadaptations involved in amphetamine and cocaine addiction. *Drug Alcohol Depend* 51:141–153.

Whitelaw RB, Markou A, Robbins TW, Everitt BJ (1996) Excitotoxic lesions of the basolateral amygdala impair the acquisition of cocaine-seeking behaviour under a second-order schedule of reinforcement. *Psychopharmacology* (Berl) 127:213–224.

Wise RA, Bozarth MA (1987) A psychomotor stimulant theory of addiction. *Psychol Rev* 94:469–492.

Wise RA, Newton P, Leeb K, Burnette B, Pocock D, Justice JB, Jr. (1995) Fluctuations in nucleus accumbens dopamine concentration during intravenous cocaine self-administration in rats. *Psychopharmacology* (Berl) 120:10–20.

Wolf ME, Sun X, Mangiavacchi S, Chao SZ (2004) Psychomotor stimulants and neuronal plasticity. *Neuropharmacology* 47 Suppl 1:61–79.

Woodward DJ, Chang J-Y, Janak P, Azarov A, and Anstrom K (1999) Mesolimbic neuronal activity across behavioral states. In: Advancing from the Ventral Striatum to the Extended Amygdala, McGinty JF Ed., *Ann. N.Y. Acad. Sci.*, New York, 91.

Wyvell CL, Berridge KC (2000) Intra-accumbens amphetamine increases the conditioned incentive salience of sucrose reward: enhancement of reward "wanting" without enhanced "liking" or response reinforcement. *J Neurosci* 20:8122–8130.

Yoshimura M, Higashi H, Fujita S, Shimoji K (1985) Selective depression of hippocampal inhibitory postsynaptic potentials and spontaneous firing by volatile anesthetics. *Brain Res* 340:363–368.

Zhang GC, Mao LM, Liu XY, Wang JQ (2007) Long-lasting up-regulation of orexin receptor type 2 protein levels in the rat nucleus accumbens after chronic cocaine administration. *J Neurochem* 103:400–407.

Zhang XF, Hu XT, White FJ (1998) Whole-cell plasticity in cocaine withdrawal: reduced sodium currents in nucleus accumbens neurons. *J Neurosci* 18:488–498.

Zhang XF, Cooper DC, White FJ (2002) Repeated cocaine treatment decreases whole-cell calcium current in rat nucleus accumbens neurons. *J Pharmacol Exp Ther* 301:1119–1125.

APPENDIX: DESCRIPTION OF TECHNIQUES

SURGERY

INTRAVENOUS CATHETER

Catheter: A small Silastic tubing (0.025 o.d.) is inserted around the inferior part of the cannula, and then a larger Silastic tubing (0.046 o.d.) is inserted to reinforce insertion area around the cannula. A bubble of silicone is placed at 3.7 cm from the extremity of the small tubing. This extremity will be inserted into the right jugular vein. The opposite end of the catheter is connected to a 22-gauge guide cannula curved at 90 degrees in the inferior part. The 22-gauge cannula tubing is then cemented inside a nylon bolt. This terminal end of the catheter exits between the scapulae and is anchored there by means of sutures and a small piece of Marlex mesh.

CATHETERIZATION

Before the start of the surgery, rats are deeply anesthetized with ketamine (30 mg/kg IP) and xylazine (5 mg/kg IP). The back and neck are clipped with electric clippers and then washed with betadine. A 1 cm incision is made throughout the skin of the neck diagonally from the mandibule to a point midway between midline and shoulder. Using forceps, the skin is separated from the muscle, and the muscle is bluntly dissected to expose the jugular vein. Using microforceps, the vein is isolated by teasing away surrounding muscle and tissue. A second incision is made in the back of the animal midway between the base of the tail and shoulders. The 22-gauge cannula tubing is placed under the skin, and the extremity of the cannula will exit in the middle of the back. Thereafter, the catheter is passed subcutaneously until it exits through the first incision, above the jugular vein. The jugular vein is cut, using Vannas scissors, and the catheter is inserted into the vein up to the bubble of silicone using a trocar. The catheter is secured to the vein with surgical silk sutures and then sutured to underlying smooth muscle. The silk suture is sterile and nonreactive and is necessary to provide a permanent anchoring of the catheter. The skin incisions are closed with individual chromic gut sutures, washed with sterile saline. Chromic gut suture is used to maximize rate of wound healing and minimize possibility of secondary injuries or infections associated with exposed suture accessible to the animal. A metal cap is placed on the extremity of the cannula to close the catheter. The wound is sprayed with gentamicin sulfate (GentavVed) topical spray.

MICROWIRE IMPLANT

Directly following the catheterization surgery, rats are placed into the stereotaxic apparatus and prepared for implant of the microwire array. The scalp incision is extended anterior to a point between the eyes. The facia is pushed back to the bone ridge, and the skull is thoroughly washed with saline. Using a dental drill, holes are drilled to accommodate six "skull screws." We have traditionally use 1-72 × 1/8 slot pan m/s stainless steel screws from JET Fitting and Supply Corporation (Santa Ana, California). It can be helpful to use smaller screws in the most anterior plates of the skull. Screw holes should be placed fairly close to the midline of the skull. This allows adequate clearance between the screws and the bone ridge to apply dental cement. A rectangular hole, consistent with the rectangular shape of the array and slightly larger than the outer dimension of the array (hole dimension = 2.2 mm × 0.6 mm) is drilled through the skull above the accumbens (Paxinos and Watson 2004). An additional hole is drilled for the ground electrode. The skull is thoroughly cleaned with saline and allowed to dry.

The skull screws are then screwed into place. The ground wire is pushed under the skull and cemented to the skull screws. Cement is applied so as to completely cover screws and wires, but care is taken to avoid any cement interfering with the array implant. Thereafter, the connector strips of the array are attached to a custom-made "array holder" that is mounted on the stereotaxic apparatus. The array is then lowered into brain slowly and in stages to avoid dimpling of brain. The array is lowered in approximately 1-mm steps over 1-min periods. Each 1-mm lowering is separated by 5 min to allow the brain to recover. During the interval between successive lowering of the array, the well is filled with saline. This melts away polyethylene glycol from the array. During the lowering procedure care is taken to avoid contact between the wires and either the bone edges of the skull hole or any other implement that might scratch the wire. Before lowering the array the last 0.2 mm, the skull hole is filled with cement. Once the final lowering is made, the cement is allowed to thoroughly dry before continuing with the surgery. The array is then encased and cemented in place using dental acrylic. It should be noted that the dental acrylic does not bond to the Teflon wires; it is thus possible, given adequate pressure, to displace the microwires (e.g., pull them back up dorsally along the electrode track), at least at the early stages of applying the acrylic and before the connector is cemented into place. It is thus optimal to complete as much of the cementing before moving the connector end of the array.

POSTOPERATIVE CARE

Immediately after surgery, animals remain on a heating pad until they begin to move. Animals are then placed in the newly cleaned stainless steel grid holding cage. The catheters are flushed daily with 0.2 ml of an ampicillin solution (0.1 g/ml) containing heparin (300 IU/ml) to maintain patency, and the animals are checked to confirm that they heal well and that they are free of signs of pain and distress. We check for the following: normal consumption of food and water, evidence of normal elimination and grooming, normal movement around the home cage, and good wound condition. We allow animals a minimum of 7 days for recovery time before any further behavioral training.

RECORDING EQUIPMENT AND PROCEDURES

MICROWIRE ARRAY

Descriptions of microwires and various microwire arrays, along with photographs and diagrams, have been provided by other authors (e.g., Moxon 1999; Woodward et al. 1999). The array that we use is manufactured by Microprobes, Inc., (Gaithersburg, Maryland) and is viewable on their Web site (www.microprobes.com). Briefly, 16 microwires and 1 ground wire are soldered into two Omnetics (Minneapolis, Minnesota) miniature connector strips (GF-8). The solder connections are coated with epoxy, and the microwires are supported with polyethylene glycol to keep their spacing. Although the materials and configuration of the arrays can be customized, we use a rectangular array that consists of two rows of 8 quad-Teflon-coated stainless steel microwires (California Fine Wire, Grover City, California). The diameter of each wire in the array is 50 µm with no insulation. The two rows of microwires are separated from each other by 0.50 mm. Adjacent wires within each row are approximately 0.25 mm apart (wire center to wire center). This array configuration yields reliable success in recording the activity of single cells. Our early experience with microwire recording indicated that the use of the array had numerous advantages, relative to a bundle of wires. First, there is evidence that the yield of usable wires is greater and that the tissue damage is reduced by use of the array, relative to the bundle. Additionally, given the geometric configuration and between-wire spacing, it is possible to ultimately identify the location of each wire tip through histological analysis (discussed further below).

RECORDING EQUIPMENT

Each microwire in the array is connected via an independent channel to a computer-controlled recording system. We use a system built by Plexon, Inc. (Dallas, Texas), although comparable systems are made by other suppliers. All of these systems contain the same basic components, which include a head-mounted headstage amplifier, an electronic harness, a swivel, and a computer-controlled amplification and filtering system. These components will be described in detail in the following sections. All systems can be viewed in detail on the Web sites of the suppliers.

TETHERING SYSTEM

The tethering system contains the tether itself, a swivel, and a counterbalance. This system protects the catheter and electronic harness, transfers rotational force to the swivel, and provides strain relief for both the harness and the catheter.

The tether length is a critical variable that must be closely attended to. It influences the freedom of movement of the animal and the proper working of the tether system. A tether that is too short prevents the animal from moving to the outer edges and corners of the operant chamber. Because animals, particularly those exposed to drugs such as cocaine, move around a lot, the interface between the animal and the catheter or microwire array can be stressed. On the other hand, a tether that is too long tends to have inadequate torque and can twist on itself rather than rotate the swivel. This can lead to damage of the tether and thus the catheter or wire connections. A tether that is too long can also loop toward the animal during rearing and thus become accessible to the animal. Tethers are commercially available, although the length may need to be customized according to the dimensions of the operant chamber.

A swivel, or commutator, is necessary to reduce strain on the implant as the animal moves around. The swivel used in our experiment is the Airflyte CAY-675-24 electronic and fluid swivel (Airflyte Electronics, Bayonne, New Jersey). This swivel has low contact noise, low rotational torque, and O-ring seals that can withstand both alcohol sterilization and salt solutions. This swivel is additionally light and compact enough to be counterbalanced (see below). A simple fluid swivel is used for behavioral training when recordings are not in progress.

The counterbalance serves to reduce strain on the tether when the animal approaches the perimeter of the chamber while also moving the tether away from the animal when it rears. This serves to both reduce stress on the implant and catheter, and also reduce the possibility of the animal damaging the tether. Counterbalances are available commercially from Med Associates (St. Albans, Vermont), although they may need to be customized to accommodate the dimensions of the operant chamber.

OPERANT CHAMBERS

The operant chambers that we use are commercially purchased from Med Associates (St Albans, Vermont). The chamber is made largely of Plexiglas. The inner walls and floor of the chamber are free of any metal or protrusions. All devices (lights and speakers) are mounted on the outside of the chamber. The floors are removable and made of Plexiglas square rods. The operant chambers are housed in sound-attenuating outer chambers. In designing the placement of operant chambers and devices such as lights, pumps, and levers, it is important to anticipate situations that can introduce noise into the recordings. If the animal makes direct contact with a metallic part of the operant chamber, current may flow between the animal and that part, whch can cause disruptive artifacts in the recording. For this reason, we use nonconductive, plastic walls and coverings on all devices that the animal has access to. Devices such as retractable levers that are operated by DC motors can also generate electromagnetic fields that can cause artifacts without direct contact between the animal and the device. We have eliminated the potential for these artifacts by carefully grounding and shielding all devices that may generate such a field. The specific source of electrical artifacts will depend on the specifics of the experimental setup. Therefore, it is important to test for any potential sources of noise and to take steps to reduce the noise before carrying out experiments.

RECORDING EQUIPMENT

Voltage signals from individual microwires are first amplified 20× on a head-mounted headstage amplifier. Amplification at this early stage is important, because the signals are very small and would be degraded if they were transferred over large distances. The headstage amplifier is connected through the tether and swivel to a preamplifier, mounted outside of the operant chamber. This preamplifier amplifies the signal another 50× and also subtracts the signal of each microwire from a user-selected reference microwire. In general, the user selects a wire in the array that does not have any neural signals on it. Subtracting this reference signal from the signal on each microwire removes common sources of noise from all wires, such as noise caused by chewing or locomotion. This processed signal is transferred to a final amplification and filtering stage. The gain of this amplification is user selectable from 1× to 32×. This is useful for fine-tuning the gain of each channel to best detect neural signals. Also at this stage, a high-pass filter is applied to remove slow, cyclic activity such as the 60Hz noise. A user-selectable threshold is also applied to the neural signal at this stage, and all voltage events smaller than this threshold are discarded. Typically this threshold is set to be slightly larger than the noise band, such that large noise events and neural signals are retained. These events are digitized and stored for later offline analysis.

DATA ANALYSIS

After the experiment is over, the digitized waveforms are sorted into groups, so that action potentials from individual neurons can be separated from large noise events. Briefly, this procedure is called "spike-sorting" and consists of grouping the voltage events into groups of similar shapes, using parameters such as amplitude of the waveform, or principal components extracted from a sample set of waveforms. Three-dimensional scatterplots are used to plot these features against each other, which allows for visual discrimination of different groups as different clusters in the three-dimensional space. In general, there is a cluster of "noise" waveforms and one or more clusters of neural waveforms. Various controls are used to reduce the possibility that multiple neurons contributed to the activity in a cluster of neural waveforms. For example, accumbal neurons have a refractory period of ~2msec. An automated procedure examines the interspike interval of all spikes and counts the percentage of spikes that occurred within 2 msec of each other. If this number is high, it may indicate that more than one neuron is contributing spikes to the cluster, considering it would be unlikely that a single neuron would fire so frequently within its own refractory period. It should be noted that these sorting procedures are far from standard, and the optimization and automation of them is the subject of ongoing research (Buzsaki 2004; Schmitzer-Torbert et al. 2005; Adamos et al. 2008; Chan et al. 2008). At the present time, commonly used methods of spike sorting involve a combination of computer-assisted and manual sorting of spikes. For additional discussion regarding data analysis, see Peoples (2003).

HISTOLOGY

At the end of the experiment, histological procedures are used to confirm the location of the tip of the recording electrodes. Subjects are injected with a lethal dose of sodium pentobarbital. Anodal current (50 µA for 4 seconds) is passed through each microwire. Animals are perfused with formalin-saline. Coronal sections (50 µm) are mounted on slides and incubated in a solution of 5% postassium ferricyanide and 10% HCl to stain the iron deposits left by the recording tip (Green 1958). The tissue is counterstained with 0.2% solution of Neutral Red. The location of each wire tip is plotted on the coronal plate (Paxinos and Watson 1998) that most closely corresponds to its anterior–posterior position. The microwire array is rectangular in configuration and made of two parallel rows of wires. The configuration and geometric relationship of the various wires to each other is usually maintained as the wires are lowered into brain so that it is possible to identify the location of each of the 12–16 wire tips during histology.

3 Neurochemistry of Addiction

Monitoring Essential Neurotransmitters of Addiction

Stefan G. Sandberg and Paul A. Garris

CONTENTS

3.1   INTRODUCTION AND SCOPE

Developments in chemical monitoring have provided the neuroscientist with a powerful toolbox for investigating biological mechanisms of drug addiction. Indeed, by sampling time-dependent changes in neurotransmitters *in situ*, these techniques have contributed prominently to our understanding of the neural substrates of goal-directed behavior and how drugs of abuse alter the functioning of these neural systems and ultimately cause addiction. Historically, chemical monitoring in the central nervous system has been dominated by voltammetry and microdialysis. Fundamentally different in analytical strategy, voltammetry and microdialysis bring distinct strengths and weaknesses to neurotransmitter measurements.

Although this chapter introduces principles of voltammetry and microdialysis and provides a brief historical perspective of these two techniques, the primary focus is newer technologies emerging in the last decade or so. Admittedly, and at the risk of bias, most coverage is devoted to recent advances in voltammetry. The reason is straightforward: While conventional microdialysis has remained the workhorse technique for this sampling approach in the modern era of chemical monitoring, the field of voltammetry has been revolutionized by the emergence of the so-called "real-time" techniques monitoring neurotransmitter with subsecond temporal resolution. Originally developed in a few laboratories, these rapid sampling techniques now have become more widely adopted and have supported diverse applications. To provide a flavor for new developments outside voltammetry, recent advances in microdialysis and in a technique related to voltammetry, enzyme-linked biosensors, are also described. Although clearly exciting and important for the future of chemical monitoring, these developments have yet to reach broad adoption beyond the laboratory of origin, and as a result, applications have been more limited to date. Hence, coverage of microdialysis and enzyme-linked biosensors is less extensive and more general than voltammetry.

To illustrate new developments in the field of chemical monitoring, applications focus on four neurotransmitters/modulators, dopamine, GABA, glutamate, and neuropeptides, established as playing essential roles in addiction (Kalivas and Volkow 2005; Wise 2005; Hyman 2005). As such, this chapter highlights the rapid sampling voltammetric techniques of amperometry, high-speed chronoamperometry and fast-scan cyclic voltammetry for monitoring dopamine. All four analytes are amenable to microdialysis, and emphasis here is placed on capillary liquid chromatography and capillary electrophoresis for separation, and laser-induced fluorescence and mass spectrometry for detection. Finally, recent advances in enzyme-linked biosensors are described for measuring glutamate. Application of these techniques to other neurotransmitters is also noted.

3.2 PRINCIPLES OF VOLTAMMETRY AND MICRODIALYSIS

Several excellent reviews are available for voltammetry (e.g., Justice 1987; Marsden et al. 1988; Stamford 1989; Adams 1990; Kawagoe et al. 1993; Blaha and Phillips 1996) and microdialysis (e.g., Benveniste and Hansen 1991; Ungerstedt 1991; Justice 1993; Westerink 1995; Watson et al. 2006). The entire list is long, and only a few are referenced above. Collectively, this literature details technical aspects, practical considerations, history of development, applications, and of course, limitations and controversies. The reader is encouraged to examine the broader literature in order to augment the necessarily more limited coverage provided here.

3.2.1 GENERAL PRINCIPLES

Four analytical characteristics are important when considering chemical monitoring in the brain: sensitivity, selectivity, speed, and size. Generally speaking, voltammetry exhibits better temporal and spatial resolution, whereas microdialysis is more sensitive and selective. Microdialysis is additionally more versatile, with a greater capability for monitoring multiple analytes and a more diverse set of analytes. Distinct analytical characteristics reflect fundamental differences in the manner in which chemical monitoring is achieved.

The superior sensitivity, selectivity and versatility of microdialysis directly result from independent analysis of a sample distal to collection at the probe. Indeed, powerful analytical methods have been coupled to microdialysis. By contrast, all chemical information regarding analyte identity and quantity is collected by the voltammetric probe itself while implanted. This strategy ultimately places restrictions on sensitivity and selectivity, as well as the chemical species amenable to detection. On the other hand, the sampling strategy of voltammetry permits fast measurements and small size, hence the superior temporal and spatial resolution.

A variety of voltammetric approaches have been developed over the years for chemical monitoring in brain tissue. All share a common principle: An electrical potential is applied to the electrode, and the resulting current is measured. Current consists of charging and faradaic components. Charging current is an unwanted signal generated by capacitance at the electrode-solution interface. The electrochemical signal of interest, faradaic current, results from analyte oxidation (i.e., loss of electrons) or reduction (i.e., gain of electrons) and is proportional to the number of molecules electrolyzed. An instrument known as a potentiostat applies voltage and concomitantly measures current. Several neurochemicals are electroactive and are thus amenable to detection by voltammetry. These include the monoamines, dopamine, norepinephrine, and serotonin and their metabolites, ascorbic acid, oxygen, hydrogen peroxide, and nitric oxide (Justice 1987; Borland and Michael 2007).

The voltammetric measurement of dopamine is conceptually shown in Figure 3.1A. When the electrode is driven to a sufficiently positive potential, the neurotransmitter oxidizes, and two electrons per molecule are transferred to the electrode. The product of dopamine oxidation, dopamine-*ortho*-quinone, is also electroactive and is reduced back to dopamine when the potential of the electrode becomes sufficiently negative. In this case, two electrons are transferred from the electrode back to the quinone during reduction. Because the electroformed quinone diffuses away from the electrode, the reductive step must occur very quickly after dopamine oxidation in order for the quinone to be measured. In brain extracellular fluid, the quinone is chemically reduced back to dopamine by ascorbic acid undergoing electrocatalytic oxidation.

Voltammetric electrodes have been fabricated from carbon paste or fibers (Blaha and Phillips 1996). The sensing surface of a carbon-paste electrode is a disk with a diameter between 150 and 300 μm. Disk electrodes made from a single carbon fiber are considerably smaller. The sensing region is an ellipse, with a minor radius the diameter of the carbon fiber (~5 to 30 um). The major radius is slightly larger and depends upon the angle of polishing. For cylindrical electrodes, the entire length of

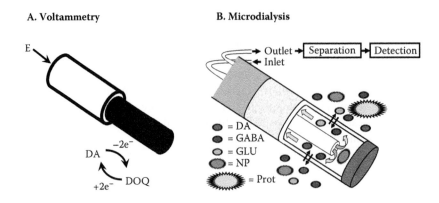

A. Voltammetry   B. Microdialysis

FIGURE 3.1 (Please see color insert following page 112.) Comparison of voltammetry (A) and microdialysis (B).

the exposed carbon fiber (~20 to 400 µm) serves as the sensing surface. A variety of waveforms have also been applied to alter electrode potential. Sweep (e.g., linear sweep voltammetry) or pulsed (e.g., chronoamperometry) methods, or a combination of both (e.g., differential pulse voltammetry), were originally used. These so-called slow voltammetric techniques measured analyte with 1-minute temporal resolution.

In microdialysis, contents of extracellular fluid are extracted by an implanted probe and physically removed from brain for chemical analysis. As schematically shown in Figure 3.1B, the microdialysis technique thus has three components: probe, means for separation, and means for detection. Different probes have been used, but a common configuration is concentric tubing covered at the implanted end by a semi-permeable membrane (Watson et al. 2006). Molecules in extracellular fluid smaller than the molecule-weight cutoff of the dialysis membrane diffuse into the probe fluid, the so-called dialysate, down a concentration gradient maintained as perfusate is pumped (0.1 to 3 µl/min) between inner and outer tubing. Artificial cerebrospinal fluid is used as perfusate to minimize dialysis of extracellular ions. Probe dimensions are typically 200 to 400 µm in diameter and 1 to 4 mm in length. The temporal resolution of conventional microdialysis is about 10 min.

A workhorse technique for chemical analysis by conventional microdialysis is reverse-phase high-performance liquid chromatography with electrochemical detection. Separation is based on hydrophobicity. An aqueous mobile phase is pumped through a column packed with a hydrophobic stationary phase containing long alkyl chains ($-(CH_2)_n-CH_3$). The more hydrophobic the analyte is, the more it interacts with the column material and the longer the retention time. Electroactive analytes are detected as they leave the column. Carbon paste and glassy carbon have been used as electrode material, and the principles of operation are the same as voltammetry. Monoamines and their metabolites are oxidized by holding the electrode at a positive potential.

Nonelectroactive neurotransmitters are detected by other means (Westerink 1995). For example, amino acid neurotransmitters, such as glutamate and GABA, are first derivatized (i.e., chemically modified by adding an electroactive or florescent label) so that they are amenable to either electrochemical or florescence detection.

Acetylcholine is detected by a two-step enzymatic process, degradation to choline by acetylcholine esterase and conversion to hydrogen peroxide by choline oxidase, followed by electrochemical detection of hydrogen peroxide at a platinum electrode. Neuropeptides are detected by radioimmunoassay, a technique that uses antibodies directed at the analyte of interest for quantification.

3.2.2 Historical Perspective

By far, the most controversial issue plaguing voltammetry is selectivity. The crux of the matter is that many electroactive neurochemicals, including the mono-amines, their metabolites, and ascorbic acid, have overlapping oxidation potentials. Moreover, the voltammetric electrode is equivalent to the electrochemical detector in microdialysis, yet no column is available to separate individual analytes prior to measurement. Given the enormous resolving capabilities of analytical columns, the limitations placed on voltammetry for selective chemical measurements are thus rather obvious. Because of these considerations, Wightman and co-workers proposed five criteria for verifying the identity of signals obtained voltammetrically: electro-chemical, anatomical, physiological, pharmacological, and independent chemical analysis (Wightman et al. 1987; Phillips and Wightman 2003).

Considerable effort was extended to improving the selectivity of the early voltam-metric techniques (Justice 1987; Adams 1990; Blaha and Phillips 1996). Electro-chemical pretreatment of the electrode enabled the resolution of three distinct signals: ascorbic acid, catechols (i.e., dopamine and its metabolites), and indoles (serotonin and its metabolites)/uric acid. Further discrimination of dopamine and serotonin proved more difficult. Another strategy was to modify the electrode chemically by the addition of selective films, such as Nafion or stearic acid. The rationale is that negative charges present in the film retard access of the negatively charged ascorbate and acid metabolites to the electrode surface, while permitting, and even enhancing, availability of the positively charged neurotransmitter. Incorporating enzymes into the electrode design, in order to remove interferents, was also employed.

Despite attempts to improve selectivity, the early era of voltammetry is perhaps best summed up by Joseph (1996): "While it is clear that naturally induced changes in [dopamine] release will show up as changes in the [electrochemical] signal, it is not yet finally established that no other substance can contribute to it for any of the [*in vivo* voltammetric] techniques." As a result, conventional microdialysis had become the dominant technique for monitoring dopamine in the brains of freely behaving animals by the early 1990s. Superior chemical resolution was ultimately the deciding factor.

As recently discussed in a review by Kennedy and co-workers (Watson et al. 2006), microdialysis is not without its pitfalls. In addition to temporal resolution, the most notable issue is damage caused by probe implantation. While precisely relating the observed anatomical trauma to physiological consequences is difficult, damaged tissue appears to distort analyte dynamics and complicate determining an absolute concentration. For example, during "no net flux," the most common method for estimating basal analyte levels, different concentrations of the analyte of interest are perfused into the probe and subsequently measured at the probe outlet. Basal level is estimated when a perfused concentration equals the outlet concentration, indicating that no analyte has diffused from or into the probe. However, the no net flux method assumes that analyte extraction from the probe equals analyte recovery from tissue, which may not be the case with damaged tissue surrounding the dialysis probe. Another issue apparently related to implantation damage is that measured levels of some neurotransmitters, such as glutamate and GABA, are not sensitive to tetrodotoxin (TTX), a blocker of voltage-gated Na^+ channels, or dependent on Ca^{2+}, as expected if originating from neurons.

Detailed in the next two sections, new developments in chemical monitoring over the last decade or so have both further enhanced the respective strengths of voltammetry and microdialysis, as well as minimized weaknesses, thus blurring traditional distinctions between applications. The "modern era" of voltammetry is distinguished by the dominance of the rapid sampling techniques of amperometry, high-speed chronoamperometry, and fast-scan cyclic voltammetry. The capability for subsecond chemical monitoring has also brought about a philosophical change toward applications. As opposed to competing with microdialysis for measuring basal analyte levels, voltammetry focused on monitoring fast signals in order to characterize neurotransmitter dynamics and mechanisms such as release, uptake, and diffusion. This type of neurobiological study is clearly a much better fit for the strengths of voltammetry: fast measurements at a small probe.

One of these rapid sampling techniques, fast-scan cyclic voltammetry, has also managed to resolve monoamines electrochemically from their metabolites and other interferents by virtue of the background-subtracted voltammogram. This development has led to the capability for monitoring naturally occurring, subsecond changes in dopamine levels associated with specific aspects of behavior. Moreover, by combining the background-subtracted cyclic voltammogram with chemometrics, it is now possible, under some conditions, for voltammetry to characterize changes in basal dopamine levels, thus moving this technique back into the domain of microdialysis.

While conventional microdialysis stills remains a workhorse technique for chemical monitoring in the modern era, sensitivity, selectivity, and versatility have been improved even further. Developments have occurred in methods for separation, such as capillary liquid chromatography and capillary electrophoresis, and for detection, such as laser-induced fluorescence and mass spectrometry. One upshot of these advancements is an increase in sensitivity, which has led to faster sampling rates. Indeed, the capability for monitoring neurotransmitter with a temporal resolution of a few seconds not only addresses one of the historical criticisms of microdialysis but also brings this technique closer to the domain of modern voltammetry. Additionally, distinct advantages that mass spectrometry brings to multianalyte determination are especially valuable for chemical analysis of the numerous neuropeptides and for identifying novel neurotransmitters.

Somewhat surprisingly, probes for voltammetry and microdialysis have changed little in practice in the last decade or so. As such, the carbon-fiber electrode and concentric tubing design are predominately used today. This is not to say that probe research is inactive. On the contrary, miniaturized probes are being developed for direct sampling or push–pull perfusion that may provide an alternative to the larger microdialysis probe and address the issue of implantation damage. Important advances have also occurred in the broad field of electrochemical sensors (Bakker and Qin 2006), to which the voltammetric electrodes belong. Another type is the biosensor, which contains a "biological recognition element" that undergoes a chemical reaction with the analyte. Electrochemical interrogation of this reaction permits monitoring of nonelectroactive analytes. As described in the final section, great strides have recently occurred in enzyme-linked biosensors for glutamate.

3.3 RECENT ADVANCES IN VOLTAMMETRY

The three most commonly used voltammetric techniques in the modern era are amperometry (AMP), high-speed chronoamperometry (HSC), and fast-scan cyclic voltammetry (FSCV). Conceptually and instrumentally, AMP is the simplest technique, because electrode potential is held constant and measured current most directly reflects changes in analyte. In contrast, potential in HSC and FSCV is altered with time, which produces a more complex current response that must be manipulated to reveal the underlying signal of interest. Differences in the manner in which potential is applied, and the resulting current measurements, provide both distinct advantages and disadvantages.

3.3.1 TECHNICAL ASPECTS

In AMP, the electrode is held constant at a potential sufficient for oxidizing or reducing analyte. As such, background current is relatively small. AMP is also an inherently fast technique; as soon as analyte interacts with the sensing surface, it is electrolyzed. Current can therefore be measured as fast as data are acquired by hardware for analog-to-digital conversion. The fast temporal response of AMP has been exploited for monitoring neurotransmitter release from individual vesicles during exocytosis (Wightman et al. 1991; Chow et al. 1992). The measured current spike, representing the vesicle emptying its chemical contents into extracellular space and monitored by a disk carbon-fiber electrode positioned immediately adjacent to the cell, exhibits a short duration of tens of milliseconds.

For the amperometric measurement shown in Figure 3.2, the carbon-fiber electrode was fixed at a potential (0.4 V) sufficient to oxidize dopamine. Each datum represents the current measured every 100 ms in response to 1 μM dopamine. This recording was collected by flow injection analysis (FIA; Kristensen et al. 1986), a common approach for evaluating sensors *in vitro*. In FIA, the microsensor is placed in a flowing stream of buffer and a bolus (i.e., a step change) of analyte is introduced at a known concentration. The amplitude of the steady-state signal is used for calibration, while the time to reach this steady state is used for determining response time. The dopamine bolus was injected between 5 and 15 s. Small deflections in the signal associated with these times are artifacts of the computer-controlled injection and can be ignored. There is also a ~2 s delay before injected dopamine reaches the downstream electrode. Nevertheless, the rapid, step increase and the squarelike shape of the signal highlight the attractiveness of AMP for capturing fast neurotransmitter dynamics.

Potential is stepped between two limits in HSC by the application of a square wave. For the measurement of dopamine shown in Figure 3.3A, the square wave

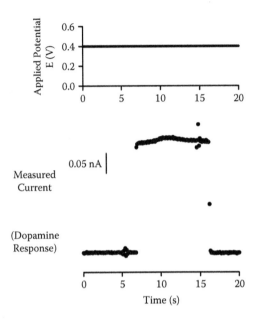

FIGURE 3.2 Amperometry.

is applied continuously between a baseline of 0 V and an oxidizing potential of 0.55 V. Each step has a duration of 100 ms, which means that dopamine is sampled every 200 ms. Because potential is stepped instantaneously, the measured current is complex. In particular, each step initially causes a large charging current that subsequently decays with time to a baseline. During the positive step to 0.55 V, current decays to a steady-state level that is comparable to the measurement in AMP.

A similar current response is measured in the presence of dopamine, except that current is slightly larger due to faradaic current (Figure 3.3B). Dopamine is oxidized to dopamine-*ortho*-quinone during the positive step, but unlike AMP, the electroformed quinone is re-reduced back to dopamine during the negative step. Current is typically integrated over the last portion of the positive step (e.g., 70 to 90 ms), when the large charging current has subsided, and is plotted every 200 ms to characterize the temporal change in dopamine levels. The FIA recording collected by HSC shares with AMP a fast, squarelike response to a dopamine bolus. Additional information is obtained with HSC by taking the ratio of currents integrated during the negative and positive steps. Because reversibility of the redox reaction differs with analyte, this ratio can be used for identification (Gerhardt and Hoffman 2001).

The potential of the microsensor is linearly scanned by applying a triangle wave during FSCV. In Figure 3.4A, top, the triangle wave begins at a baseline potential of −0.4 V, peaks at 1.0 V, and returns to the initial potential. Because potential is ramped quickly (300 V/s), the entire scan occurs in ~9.3 ms. The electrode rests at the baseline potential during the time between scans. Scans are applied at 100 ms intervals, according to the rule of thumb that ~10 times the scan duration is required to relax diffusion gradients generated during the scan. Like HSC, the rapid change in potential causes a large charging current. However, because potential is continuously

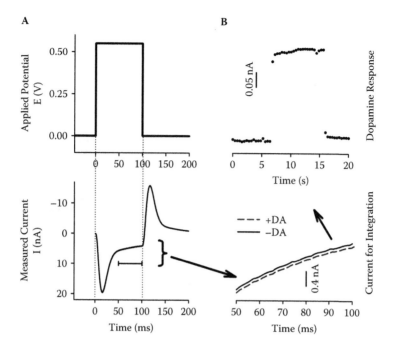

FIGURE 3.3 High-speed chronoamperometry.

changing during FSCV, a large charging current is measured throughout the scan, switching polarity during the second half of the triangle wave when applied potential returns to baseline. Also like HSC, charging current masks the smaller faradaic current resulting from dopamine.

Background subtraction is employed to resolve faradaic from charging current. This procedure is successful because the background signal is relatively constant, at least over short time scales, and thus can be mathematically subtracted. As shown in Figure 3.4A, bottom, two distinct peaks emerge for dopamine after background subtraction. The first occurs during the positive sweep at around 0.55 V and is due to the oxidation of dopamine to dopamine-*ortho*-quinone. The second, in the opposite direction, occurs during the negative sweep at around −0.2 V and results from reduction of the electroformed quinone back to dopamine.

Background-subtracted signals collected for each scan are viewed in time using a three-dimensional, pseudo-color plot to obtain the entire electrochemical record for a dynamic response. The x, y, and z axes are time, applied potential, and measured current, respectively. For the FIA example shown in Figure 3.4B, middle, brown color represents the background current zeroed by the subtraction procedure. Shortly after the dopamine bolus is injected at 5 s, two distinct features appear due to dopamine oxidation at 0.55 V (green-purple) and quinone reduction at 0.2 V (black-yellow). These features, which correspond to the two peaks in the background-subtracted current in Figure 3.4A, bottom, last for the 10-s duration of the bolus injection and then fade into the background brown color.

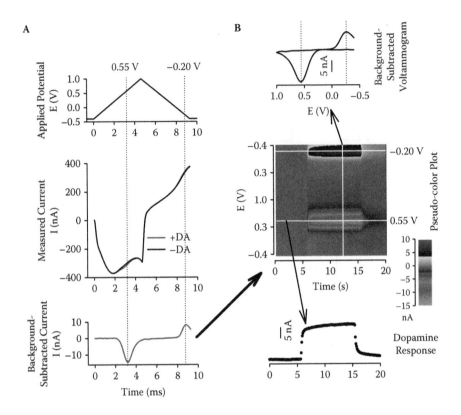

FIGURE 3.4 (Please see color insert following page 112.) Fast-scan cyclic voltammetry. (With permission from Boyd BW, Witowski SR, Kennedy RT (2000) Trace-level amino acid analysis by capillary liquid chromatography and application to *in vivo* microdialysis sampling with 10-s temporal resolution. *Anal Chem* 72: 865–871.)

Quantitative information is derived from the pseudo-color plot. The temporal response, obtained by plotting current measured at the peak oxidation potential for dopamine (0.55 V) with time (horizontal white line), is shown in Figure 3.4B, bottom. Because scans are applied every 100 ms, dopamine is sampled at this interval. Compared to AMP and HSC, the FIA response collected by FSCV is somewhat slower, with rounder edges. A background-subtracted cyclic voltammogram is generated by plotting the measured current against the applied potential (vertical white line). Shown in Figure 3.4B, top, this current versus potential relationship serves as the chemical signature identifying the analyte as dopamine. Of the known electroactive substances in brain extracellular fluid, only norepinephrine yields a similar voltammogram (Baur et al. 1988).

3.3.2 GENERAL COMPARISONS

While all three voltammetric techniques are capable of monitoring analyte with sub-second temporal resolution, AMP is clearly the fastest. AMP is typically performed with 0.1 to 16.7 ms resolution (Dugast et al. 1994; Forster and Blaha 2003; Venton et al. 2003b), which is more than sufficient for capturing extracellular neurotransmitter dynamics in terminal fields with high fidelity, even for the fastest signals. By comparison, HSC and FSCV are limited by the applied waveform and sample about an order of magnitude slower, 200 and 100 ms, respectively (Kawagoe et al. 1993; Gerhardt and Hoffman 2001).

AMP shares with HSC another advantage that is revealed by FIA, fast response time. By comparison, the dopamine bolus recorded by FSCV is temporally distorted. The source of this distortion is rooted in the measurement itself: dopamine adsorbs to the carbon surface in the time between scans (Bath et al. 2000). Consequently, fast changes in dopamine concentration are "filtered" by the slower adsorption kinetics. By oxidizing dopamine immediately upon contact, the constant potential prevents adsorption in AMP. Any adsorption during the negative step in HSC is also oxidized during the initial positive step, such that dopamine measurements later in time act similarly to AMP. In theory, the temporal distortion in FSCV is removed by deconvolution (Venton et al. 2002). The downside is that this mathematical manipulation, which restores high-frequency components filtered out by adsorption kinetics, adds noise.

Although providing the basis for superior speed, the fixed potential of AMP limits chemical resolution. With a measurement of current at only one potential, there is little electrochemical information to identify the analyte. By contrast, addition of the negative step in HSC enables the determination of a ratio for reductive to oxidative current to aid in identification. Even more powerful is the background-subtracted voltammogram of FSCV, which describes a current profile across the range of potentials for oxidation and reduction. In addition to improving identification, the negative-going potential in HSC and FSCV also reduces dopamine-*ortho*-quinone back to dopamine. Not only is the buildup of the chemically reactive quinone lessened, but dopamine levels are minimally perturbed during the measurement. By contrast, without regeneration of dopamine, a concentration gradient extending from the electrode is generated by AMP. This gradient makes postcalibration, which is used to convert current measured in brain tissue into concentration, difficult (Kawagoe and Wightman 1994; Venton et al. 2002).

Rapid scanning of the potential during FSCV is not without drawbacks. During the scan, chemical groups on the carbon surface undergo redox reactions, leading to the distinct waves on the background signal (Figure 3.4, middle). The position of these waves is sensitive to background drift and to the ions H^+ and Ca^{2+}, whose brain levels are altered by neuronal activation (Jones et al. 1995; Rice and Nicholson 1995). The response of FSCV to pH is shown in Figure 3.5A. Note that the broad features in the pseudo-color plot and background-subtracted voltammogram overlap dopamine redox potentials. Thus, monitoring current at 0.55 V results in a mixed dopamine–pH temporal response if both analytes change simultaneously. By comparison, AMP is insensitive to the same pH change, because potential is held constant (Figure 3.5B). The pH response of HSC is more complex (Figure 3.5C). HSC is quite sensitive during the early charging current phase, but insensitive when current is integrated over the last 50 ms. A small pH response was reported when current was integrated over the last 80 ms (Gerhardt and Hoffman 2001), perhaps due to the longer integration time.

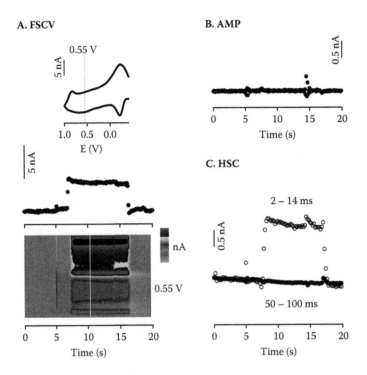

FIGURE 3.5 (Please see color insert following page 112.) Effects of pH on fast-scan cyclic voltammetry (A; FSCV), amperometry (B; AMP) and high-speed chronoamperometry (C; HCS).

Two strategies have been developed to address problems posed to FSCV by pH. The differential subtraction method (Jones et al. 1995; Venton et al. 2003a) takes advantage of the fact that pH also appears in the electrochemical record at different potentials than dopamine. Recorded at a potential distinct from dopamine, pH is simply subtracted from the mixed dopamine–pH response revealing dopamine. The limitation is that selectivity of the dopamine signal is not always established. Chemometrics has been used to address this issue. With a statistical method called principal component analysis, it is possible to resolve a pure dopamine signal from the mixed response (Heien et al. 2004). However, because of background drift, principal component analysis is restricted to signals of a limited duration. It should also be noted that differential subtraction and principal component analysis allow FSCV to monitor multiple analytes, such as dopamine, pH, and oxygen, simultaneously (Venton et al. 2003a).

Surprisingly, direct comparisons of sensitivity between the three voltammetric techniques are not readily available. Differences in electrode fabrication, instrumentation, and electrochemical parameters, and issues related to calibration, also hinder comparisons between studies. Based on *in vivo* recordings of electrically evoked dopamine levels measured at the same electrode, AMP was found to exhibit a slightly greater signal-to-noise ratio than FSCV (Venton et al. 2003b). The sensitivity of FSCV is improved by extending scan potential. For the range of –0.6 to 1.4 V, sensitivity is increased nearly 10-fold (Heien et al. 2003). However, chemical resolution and response time are reduced. For this reason, a less severe extended scan, −0.4 to 1.3 V, is also used. Detection limits for dopamine of ~15 nM have been reported using extended scan waveforms (Cheer et al. 2004; Stuber et al. 2005b), similar to that described for HSC (Gerhardt and Hoffman 2001). These detection limits are notable in that they approach the lower estimates of basal dopamine levels (~6 nM) in the striatum as determined by no net flux (Justice 1993).

3.3.3 APPLICATIONS

To the nonvoltammetrist, differences between AMP, HSC, and FSCV may appear subtle, and the contention that has at times emerged during the development of these techniques, difficult to understand. On one hand, AMP, HSC, and FSCV share the common characteristic of real-time measurement. For some applications, the techniques are therefore interchangeable, and similar results are expected and have been obtained. Nevertheless, unique advantages and disadvantages require careful consideration when selecting a technique.

General guidelines can be proposed. If analyte source is well characterized, all three techniques provide valuable information about neurotransmitter dynamics. Distinct information may, in fact, reflect more the mathematical model used for data analysis than the voltammetric technique. AMP is required for the fastest signals (<100 ms). To obtain signals devoid of distortion by ions or adsorption, AMP or HSC is most appropriate. In this case, selection of the electrode becomes critical. Nafion, used to improve selectivity and sensitivity, slows temporal response (Kawagoe et al. 1992), and larger electrodes may cause complications from tissue damage. On the other hand, if analyte source is unknown, the chemical resolution provided by FSCV is necessary. Techniques can also be combined to exploit different analytical strengths; for example, using FSCV to identify dopamine, and then, at the same electrode and recording location, switching to AMP to obtain better temporal information (Schmitz et al. 2001; Venton et al. 2003b).

Two general approaches have been developed to provide a well-characterized analyte source. The first is application of exogenous neurotransmitter by pressure ejection or iontophoresis (Gerhardt and Palmer 1987; Rice and Nicholson 1995; Kiyatkin et al. 2000). The second is evoking endogenous neurotransmitter release by electrical stimulation (Millar et al. 1985), depolarizing K^+ solution (Gerhardt et al. 1995), pharmacological agents (Suaud-Chagny et al. 1992), or visual stimuli (Dommett et al. 2005). Exogenous application is the most definitive analyte source. Recording location must be considered when evoking neurotransmitter release. Monitoring evoked dopamine levels in the richly dopamine-innervated striatum is now well established for AMP (Chergui et al. 1994; Gonon 1995; Lee et al. 2006), HSC (Gerhardt et al. 1995; Cass and Manning 1999), and FSCV (Wightman et al. 1988; Young and Michael 1993; Walker et al. 2000). Recording in mixed monoaminergic regions, such as the cortex (Mitchell et al. 1994), thalamus (Ghasemzadeh et al. 1993), amygdala (Jones et al. 1994), and midbrain (Rice et al. 1997), is more challenging.

HSC is typically coupled to pressure ejection for characterizing neurotransmitter uptake in the anesthetized rat. All three monoamines, dopamine (Gulley et al. 2007), serotonin (Daws and Toney 2007), and norepinephrine (Gerhardt 1995), have been examined in this way. Uptake determines the amplitude and clearance rate of the signal recorded at the electrode positioned near the ejection pipette (Zahniser et al. 1998). Signals can also be analyzed by Michaelis–Menten kinetics (Zahniser et al. 1999) and a model incorporating first-order uptake and diffusion (Nicholson 1985; Chen and Budygin 2007). FSCV and AMP have been applied extensively for characterizing dopamine dynamics in anesthetized rats as well, but are more commonly coupled to electrical stimulation (Millar et al. 1985; Chergui et al. 1994). The evoked signal, which increases during the stimulus train and rapidly decays to baseline afterward, contains information about release, uptake, and diffusion (Wightman et al. 1988; Schonfuss et al. 2001; Wu et al. 2001b; Venton et al. 2003b).

Applications of AMP, HSC, and FSCV in anesthetized rats have mainly provided a similar characterization of dopamine uptake in terms of regional differences (Cass et al. 1993; Garris and Wightman 1994; Zahniser et al. 1998), inhibition by psychostimulants (Cass et al. 1992; Suaud-Chagny et al. 1995; Wu et al. 2001a), and autoreceptor regulation (Dickinson et al. 1999; Benoit-Marand et al. 2001; Wu et al. 2002). However, two notable exceptions deserve mention. The first is that basal rates for dopamine uptake are slower when measured by HSC coupled to pressure ejection compared with FSCV coupled to electrical stimulation (Zahniser et al. 1999; Michael et al. 2005; Chen and Budygin 2007). The second is the demonstration by HSC and pressure ejection that uptake inhibitors, under some conditions, increase dopamine uptake rate (Zahniser et al. 1999), which has not been observed with FSCV and electrical stimulation. However, discrepancies appear related to analyte source as opposed to voltammetric technique.

Although both AMP (Falkenburger et al. 2001; Schmitz et al. 2001) and HSC (Hoffman et al. 1998) have been applied to brain slices to some extent, FSCV coupled to electrical stimulation is used extensively in this preparation for characterizing dynamics for dopamine (Palij et al. 1990; Cragg et al. 1997), serotonin (Davidson and Stamford 1995; Bunin and Wightman 1998) and norepinephrine (Palij and Stamford 1992; Callado and Stamford 2000). While information obtained regarding neurotransmitter release and uptake is similar to *in vivo* FSCV recordings, the *in vitro* approach has higher throughput and offers greater control of the extracellular milieu and drug application. *In vitro* voltammetry has factored prominently in characterizing mechanisms of abused drugs. For example, this approach has recently demonstrated that nicotine enhances reward-related dopamine signaling by increasing phasic but inhibiting tonic dopamine release (Rice and Cragg 2004; Zhang and Sulzer 2004). The three main cellular effects of amphetamine on dopamine signaling are also readily observed in the evoked voltammetric trace (Jones et al. 1998, 1999; Schmitz et al. 2001): reduced uptake by competitive inhibition, decreased exocytotic release, and increased nonexocytotic release via reverse transport.

One of the most exciting applications of rapid sampling voltammetry in the last decade is chemical monitoring in ambulatory animals. Such measurements are important for assessing the validity of measurements collected under anesthesia or in brain slices. Moreover, fast chemical monitoring in freely moving animals affords the unique opportunity for correlating neurotransmitter changes with specific aspects of behavior in real time. To date, all three rapid sampling techniques, AMP (Yavich and Tiihonen 2000a), HSC (Kiyatkin et al. 1993; Gerhardt et al. 1999), and FSCV (Garris et al. 1997; Rebec et al. 1997; Kiyatkin et al. 2000), have been applied to freely behaving animals.

Comparable measurements in anesthetized and ambulatory animals indicate that some anesthetics alter rates for dopamine uptake determined voltammetrically (Garris et al. 1997; Gerhardt et al. 1999; Sabeti et al. 2003a). Electrically evoked dopamine release is also suppressed with Equithesin, a mixture including chloral hydrate, pentobarbital, and ethanol (Garris et al. 1997). While drug mechanism is not altered, urethane temporally delays the effects of nomifensine (a psychostimulant) and haloperidol (a mixed D1/D2 antagonist) on electrically evoked dopamine levels (Garris et al. 2003).

The effects of pyschostimulants and haloperidol on dopamine levels evoked by electrical stimulation and measured by FSCV show a particularly close relationship with behavioral activation and catalepsy, respectively (Budygin et al. 2000; Garris et al. 2003; Robinson and Wightman 2004). The effects of cocaine on exogenously applied dopamine measured by HSC and locomotor activity similarly correlate (Sabeti et al. 2002, 2003b). In these studies, baseline and cocaine-induced changes in dopamine uptake measured voltammetrically were additionally linked to the behavioral responsiveness of individual rats to both acute and chronic cocaine administration. Taken together, these studies strongly suggest that indexes of dopamine signaling measured by HSC coupled to pressure ejection and FSCV coupled to electrical stimulation are behaviorally relevant.

The inherent speed of AMP, HSC, and FSCV has provided insight into the study of intracranial self-stimulation. In this classic paradigm, animals work to obtain electrical stimulation at an electrode implanted to activate neural systems involved in reward-related behavior (Olds and Milner 1954). The fast-sampling voltammetric techniques permit dopamine measurements to be linked to individual operant responses, demonstrating a dissociation between dopamine release and the reinforcing electrical stimulation under some conditions (Garris et al. 1999; Kilpatrick et al. 2000; Yavich and Tiihonen 2000b).

More recently, FSCV has been used to characterize naturally occurring dopamine signals associated with behavior. Salient stimuli elicit burst firing in midbrain dopaminergic neurons (Overton and Clark 1997; Schultz 1998) that generates a rapid, dopamine concentration spike in terminal fields (Carelli and Wightman 2004). These so-called dopamine transients typically exhibit an amplitude in the nanomolar range with a duration of a few hundred milliseconds and occur with a baseline frequency of ~1 to 2 per minute (Robinson and Wightman 2006). Dopamine transients have been linked to specific aspects of behavior during social interaction (Robinson et al. 2001, 2002), cocaine self-administration (Phillips et al. 2003), and food reward (Roitman et al. 2004) and are augmented by noncontingent administration of psychostimulants (Robinson and Wightman 2004; Stuber et al. 2005a). Spontaneously occurring dopamine transients are also observed (Robinson et al. 2002; Cheer et al. 2004).

Figure 3.6 illustrates the FSCV measurement of dopamine changes associated with phasic signaling in a freely behaving rat. This animal was administered

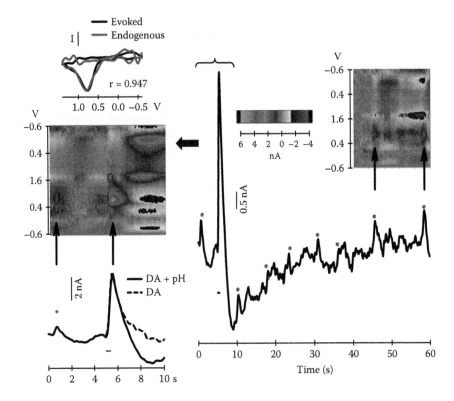

FIGURE 3.6 (Please see color insert following page 112.) Recording dopamine transients in the freely behaving rat by fast-scan cyclic voltammetry. Pseudo-color plots are on identical time scales as the recording immediately below. Electrical stimulation is denoted by the solid line under each recording. The inset shows background-subtracted voltammograms for electrically evoked dopamine release (evoked, black line) and for a naturally occurring dopamine transient (endogenous, red line). Abbreviations: DA, dopamine.

WIN55,212-2, a cannabinoid CB1 agonist, to increase the frequency of dopamine transients (Cheer et al. 2004) from about 1 to >7 per minute. The main panel shows a 60-s recording collected in the caudate-putamen. The large increase in signal near the beginning is evoked dopamine release due to electrical stimulation of the substantia nigra. Other, smaller peaks designated by an asterisk are dopamine transients. To the left is an expanded view of the first 10 s with a dopamine transient occurring between 0 and 2 s and the electrically evoked signal beginning at 5 s. Dopamine features for both signals are displayed in the pseudo-color plot in Figure 3.6.

The background-subtracted voltammogram collected during stimulation is used as a reference dopamine signal to identify transients. The inset above the pseudo-color plot shows that voltammograms collected during stimulation and for the transient correlate well. Each voltammogram collected in the recording is statistically compared with the evoked dopamine signal, and only those above a preselected threshold are considered dopamine. Other prominent features occur in the pseudo-color plot after stimulation and are due to changes in extracellular pH (left panel).

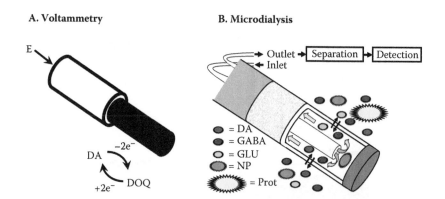

FIGURE 3.1 Comparison of voltammetry (A) and microdialysis (B).

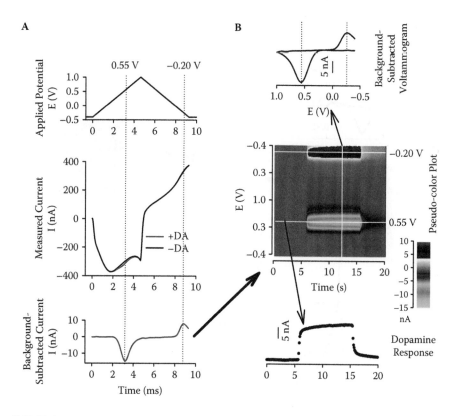

FIGURE 3.4 Fast-scan cyclic voltammetry. (With permission from Boyd BW, Witowski SR, Kennedy RT (2000) Trace-level amino acid analysis by capillary liquid chromatography and application to *in vivo* microdialysis sampling with 10-s temporal resolution. *Anal Chem* 72: 865–871.)

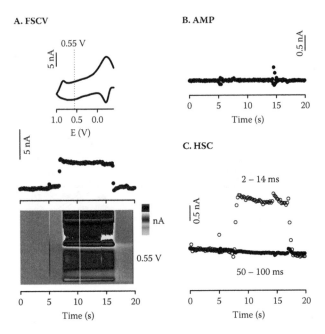

FIGURE 3.5 Effects of pH on fast-scan cyclic voltammetry (A; FSCV), amperometry (B; AMP) and high-speed chronoamperometry (C; HCS).

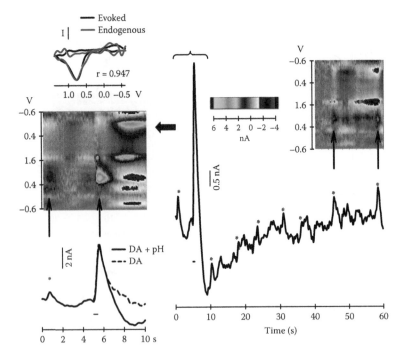

FIGURE 3.6 Recording dopamine transients in the freely behaving rat by fast-scan cyclic voltammetry. Pseudo-color plots are on identical time scales as the recording immediately below. Electrical stimulation is denoted by the solid line under each recording. The inset shows background-subtracted voltammograms for electrically evoked dopamine release (evoked, black line) and for a naturally occurring dopamine transient (endogenous, red line). Abbreviations: DA, dopamine.

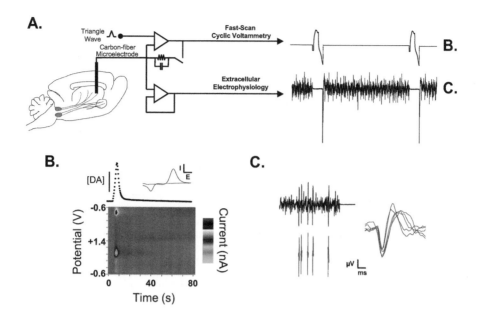

FIGURE 3.7 Quasi-simultaneous voltammetry and electrophysiology. A simplified circuit diagram is described in A. Monitoring of the electrically evoked signal by fast-scan cyclic voltammetry is shown in B. Identification of two distinct units from the electrophysiological record is shown in C.

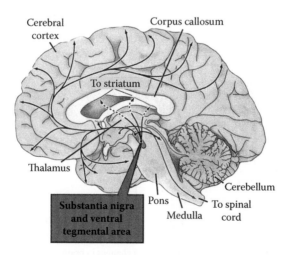

FIGURE 6.1 The distribution of dopaminergic neurons and their projections in the human brain. The cell bodies of dopamine-containing neurons exist in clusters of cells (nuclei) located in the midbrain, and their axons project to widespread regions across the brain. Dopaminergic projection to cerebral cortex is heaviest in the frontal lobe, whereas projection to the occipital cortex is sparse. Subcortical structures such as the striatum and thalamus are also prominent dopaminergic targets. Adapted from Purves, D., Augustine. H. J., Fitzpatrick, D., Hall, W. C., LaMantia, A., McNamara, J. O., and Williams, S. M. 2004 *Neuroscience, Third Edition*. Sunderland, MA: Sinauer Associates. With permission.

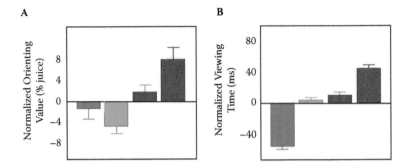

FIGURE 6.2 Monkeys value social and reproductive information. (a) Mean normalized orienting values for images of familiar monkeys. Orienting values are significantly higher for both the perinea (red bar) and high-status faces (blue bar), in contrast to either the low-status faces (green bar) or gray square (gray bar). Orienting value is measured as the amount of juice required to induce the monkey to choose equally between the image and nonimage target. Therefore, positive numbers reflect the amount of juice the subject monkey will "pay" to see the image; negative numbers reflect the amount of juice required to "bribe" the monkey to see the image. (b) Normalized-looking times for various image classes. Although the monkeys choose to orient more frequently to the high-status faces than the low-status faces, the lengths of time they gaze at both of these image classes are shorter than the time they spend viewing the perinea. Adapted from Deaner, R. O., Khera, A. V., and Platt, M. L. 2005 Monkeys pay per view: adaptive valuation of social images by rhesus macaques. *Curr Biol* 15, 543–8. With permission.

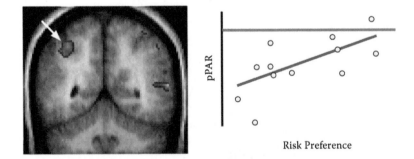

FIGURE 6.6 Subjective economic preferences predict changes in brain activation. A correlation exists between cortical activity in the posterior parietal cortex (see left, white arrow) and behavioral preference for risk. Right, each subject's mean preference value and fMRI parameter value (a measure of relative brain activity) is indicated by a single circle. Correlation = -0.62, p < 0.00001. Figure adapted from Huettel, S. A., Stowe, C. J., Gordon, E. M., Warner, B. T., and Platt, M. L. 2006 Neural signatures of economic preferences for risk and ambiguity. *Neuron* 49, 765–775. With permission.

This distortion, which causes the signal to fall precipitously after stimulation (solid line), is removed by differential subtraction, revealing the evoked dopamine signal (dashed line).

The capability to resolve dopamine transients from other signals in the electrochemical record attributed to, for example, nondopamine analytes, movement artifact, and noise, clearly highlights the utility of the background-subtracted voltammogram for chemical analysis of phasic neurotransmisson during behavior. The question arises whether FSCV can, as well, selectively monitor dopaminergic tone, the basal level of brain dopamine. Although determination of absolute concentration is still not possible, principal component analysis resolves a slow increase in steady-state dopamine levels following noncontingent administration of several drugs of abuse, including cocaine, ethanol, and nicotine (Heien et al. 2005; Cheer et al. 2007b). FSCV has additionally been used to assess decreases in the baseline voltammetric signal following local infusion of the ionotropic glutamate receptor antagonist, kynurenate (Kulagina et al. 2001; Borland and Michael 2004). Interestingly, the kynurenate-induced drop in signal represents a ~2 μM decrease in dopamine concentration, if the entire signal is attributed to this neurotransmitter.

Information about how target cells respond to measured neurotransmitter dynamics is also desirable. Separate electrodes have been used to record electrically evoked dopamine levels with AMP and extracellular single unit activity in the striatum of anesthetized rats (Gonon and Sundstrom 1996; Gonon 1997). FSCV lends itself nicely to combined chemical and electrophysiological assessment, because the carbon-fiber electrode is an excellent sensor for unit recording (Millar et al. 1981; Millar and Williams 1988) and the electrochemical measurement is not continuous. Indeed, both voltammetric and electrophysiological measurements have been made at the same carbon-fiber electrode using what is called a time-share or quasi-simultaneous procedure. Originally developed for anesthetized (Williams and Millar 1990a, 1990b) and slice preparations (Stamford et al. 1993), combined measurements have recently been made in ambulatory rats (Cheer et al. 2005, 2007a).

Figure 3.7 shows the technique of quasi-simultaneous voltammetry and electrophysiology. FSCV is performed as described above, but in the time between scans, the carbon-fiber electrode is switched to a circuit for recording units. After data manipulation, the temporal response of dopamine and histogram of unit activity are overlaid, as shown in Figure 3.8A. Data were collected in the striatum of an anesthetized rat by a carbon-fiber electrode prepared from a five-barrel micropipette, as shown in Figure 3.8B. Additional barrels were used for iontophoresis. Excitation of a glutamate-sensitive single unit by an electrically evoked signal mimicking the dynamics and amplitude of a dopamine transient is demonstrated.

3.4   RECENT ADVANCES IN MICRODIALYSIS

Unlike voltammetry, where AMP, HSC, and FSCV largely displaced the early techniques in the last decade or so, conventional microdialysis continued to be widely used for neurotransmitter monitoring, particularly in freely moving animals. Perhaps as a result, advances and new applications were not on the same large scale as those for voltammetry and described in the previous section. Nevertheless, significant

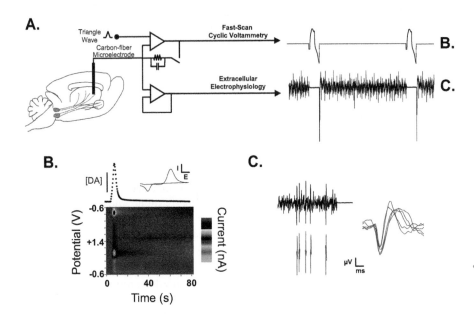

FIGURE 3.7 (Please see color insert following page 112.) Quasi-simultaneous voltammetry and electrophysiology. A simplified circuit diagram is described in A. Monitoring of the electrically evoked signal by fast-scan cyclic voltammetry is shown in B. Identification of two distinct units from the electrophysiological record is shown in C.

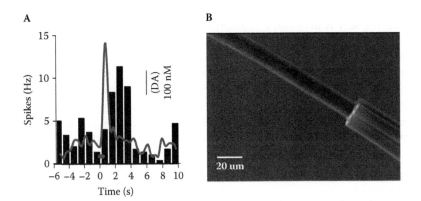

FIGURE 3.8 Coupling microiontophoresis to quasi-simultaneous voltammetry and electrophysiology. Effects of electrically evoked dopamine levels on a striatal unit excited by iontophoretically applied glutamate are shown in A. An electron micrograph of a carbon-fiber electrode fabricated in one barrel of a five-barrel micropipette is shown in B.

improvements have been made. For example, considerable effort has been directed at increasing temporal resolution, which for conventional microdialysis is about 10 min. Poor temporal resolution is due to the combination of low perfusion rates through the probe and the large sample volume requirement for high-performance liquid chromatography. Consequently, samples must be collected for long periods of time. However, the emergence of capillary liquid chromatography (CLC) and capillary electrophoresis (CE), with their high separation efficiency and low sample volume requirements, coupled to high sensitivity detection methods, such as laser induced fluorescence (LIF) and mass spectrometry (MS), have afforded faster sampling, even down to a few seconds.

3.4.1 CAPILLARY LIQUID CHROMATOGRAPHY

CLC is similar to conventional high-performance liquid chromatography in that analytes are separated based on mass exchange between the mobile and stationary phase in the separation column. CLC columns, which have an inner diameter of ~25 to 50 μm, can be viewed as a hybrid between the conventional packed column and the much smaller open tubular capillary columns, which do not contain packing material (Takeuchi 2005). Offering excellent selectivity, CLC must be coupled to high-sensitivity detection techniques because of the required low flow rates. Figure 3.9 shows the separation of amino acid neurotransmitters by CLC coupled to electrochemical detection (Boyd et al. 2000). The nonelectroactive analytes were derivatized by *o*-phthalaldehyde/*tert*-butyl thiol and monitored electrochemically with AMP by a carbon-fiber electrode. A detection limit of about 50 attomoles enabled the *in vivo* monitoring of glutamate, aspartate, GABA, and glycine with 10-s temporal resolution using a conventional microdialysis

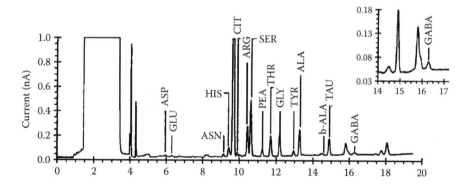

FIGURE 3.9 Chromatogram from capillary liquid chromatography with electrochemical detection. Dialysate was collected in the striatum of an anesthetized rat. The detector was a carbon-fiber electrode held at constant potential. Abbreviations: GLU, glutamate; ASP, aspartate; GLY, glycine; see also Boyd et al. (2000). (With permission from Boyd BW, Witowski SR, Kennedy RT (2000) Trace-level amino acid analysis by capillary liquid chromatography and application to *in vivo* microdialysis sampling with 10-s temporal resolution. *Anal Chem* 72: 865–871.)

probe. The neuropeptide met-enkephalin is directly electroactive and has also been detected in dialysate with CLC coupled to a carbon-fiber electrode (Shen et al. 1997).

3.4.2 CAPILLARY ELECTROPHORESIS

As shown in Figure 3.10A, CE separates analytes based on their ability to migrate in an electrical field. Since no mass exchange between mobile phase and capillary wall occurs, migration velocity is determined by the charge density and size of analyte. The mobile phase, or in CE referred to as background electrolyte (BGE), is polar and moves through the separation capillary from the negative to positive pole. This flow of BGE is referred to as electroosmotic flow and is manipulated by changing the voltage of the electric field across the capillary or the pH of the BGE. One advantage of analytes being carried by electroosmotic flow is the near absence of peak broadening, which contributes to CE's high resolution. Indeed, of the liquid separation techniques, CE possesses the highest resolving power (Issaq 1999). Due to its high separation efficiency and resolving power, CE also exhibits the fastest separation times, with analyte migrating in seconds as opposed to eluting in minutes (Issaq 1999; Moini 2002; Powell and Ewing 2005). Compared to CLC, CE supports smaller sample volumes, but this also leads to lower concentration sensitivity (Hernandez-Borges et al. 2004; Chiu et al. 2006). Mass sensitivity is largely independent of separation method.

3.4.3 LASER-INDUCED FLUORESCENCE

As conceptually shown in Figure 3.10B, in LIF, a laser beam excites analyte to a higher energy state, and fluorescence is emitted when excited analyte decays back

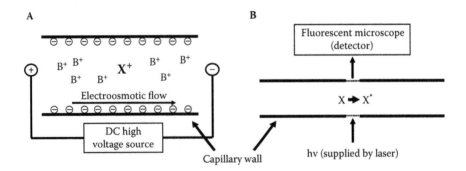

FIGURE 3.10 A. Capillary electrophoresis. Analyte X^+ is carried by electrostatic attraction and electroosmotic flow to the negative end of the capillary. B. Laser-induced fluorescence. Analyte is excited by the energy quantum supplied by a laser. Subsequent relaxation of analyte to a lower energy state results in the emission of fluorescence, which is detected by a fluorescent microscope. Abbreviations: B^+, positively charged background analyte; DC, direct current; hv, energy quanta supplied by laser beam; X^+, positively charged analyte; X^*, excited analyte.

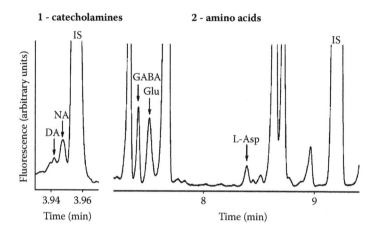

FIGURE 3.11 Electropherogram from capillary electrophoresis with laser-induced fluorescence. Dialysate was obtained from the thalamus of a freely moving rat. Abbreviations: DA, dopamine; NA, norepinephrine; Glu, glutamate; Asp, aspartate; see also Parrot et al. (2004). (With permission from Parrot S, Sauvinet V, Riban V, Depaulis A, Renaud B, Denoroy L (2004.) High temporal resolution for *in vivo* monitoring of neurotransmitters in awake epileptic cat using brain microdialysis and capillary electrophoresis with laser-induced fluorescence detection. *J Neurosci* 140: 29–38.)

to a lower energy state. One advantage of LIF is that analytes do not need to be electroactive. Because many analytes of interest do not exhibit native fluorescence or have low quantum yields, derivatization with a fluorescent label is necessary. A variety of derivatizing agents and lasers are described in the literature (e.g., Chiu et al. 2006; Garcia-Campana et al. 2007). Chen et al., using 5-furoylquinoline-3-carboxaldehyde, assayed 16 biogenic amine and amino acids including Glu, GABA, and DA (Chen et al. 2001), illustrating the importance of derivatizing agent choice. The three main ways of introducing derivatizing agents online are pre-, on-, or post-separation capillary (Garcia-Campana et al. 2007).

LIF is a highly sensitive method suitable for detecting analytes that are sufficiently separated. CE has been coupled to LIF for *in vivo* measurements of biogenic amines and amino acids in dialysate with a temporal resolution of between 3 and 100 s (Lada and Kennedy 1996; Tucci et al. 1997; Rebec et al. 2005; Shou et al. 2006). Figure 3.11 shows the CE–LIF monitoring of dopamine, norepinephrine, glutamate, aspartate, and GABA. CE–LIF has also been applied to the freely behaving rat for monitoring GABA and glutamate with 14 s temporal resolution in the nucleus accumbens following presentation of fox odor (Venton et al. 2006a) and in the basolateral amygdala during a conditioned fear paradigm (Venton et al. 2006b).

3.4.4 MASS SPECTROMETRY

The analytical technique of MS is complex, and the reader is encouraged to consult other, more detailed sources (e.g., Niessen 2003) than the very simplified coverage found in this chapter. MS is a unique detection method, because in addition to

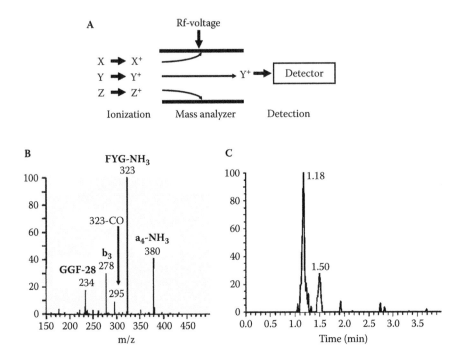

FIGURE 3.12 Mass spectrometry. Principles for a single quadrupole are shown in A. A quadrupole ion trap mass spectrogram of characteristic ion fragments from leu-enkephalin is shown in B. A reconstructed ion chromatogram, which is obtained by monitoring unique ion fragments over time, is shown for met- (first peak) and leu-enkephalin (second peak) in C. Dialysate was collected from the striatum of an anesthetized rat. Abbreviations: Rf, radio frequency; X, Y and Z, hypothetical analytes and their respective ionized state (X$^+$, Y$^+$ and Z$^+$). (With permission from Baseski HM, Watson CJ, Cellar NA, Shackman JG, Kennedy RT (2005) Capillary liquid chromatography with MS3 for the determination of enkephalins in microdialysis samples from the striatum of anesthetized and freely-moving rats. *J Mass Spectrom* 40: 146-153.)

quantification, structural information about the analyte is also obtained. Thus, when adequate separation is not reached or analytes are unknown, MS is preferred. As conceptually shown in Figure 3.12A, there are three steps to MS: ionization, mass analysis, and detection. Diverse MS configurations, with different levels of analysis capabilities, are available. Ionization strategy depends on analyte polarity and size, and on state (Thurman et al. 2001). An atmospheric pressure ionization method called electrospray ionization (ESI) is used for *in vivo* monitoring (Hernandez-Borges et al. 2004; Haskins et al. 2004; Powell and Ewing 2005). Suitable for liquid samples, ESI is a soft-ionization procedure that is compatible with almost the entire ionization continuum, including nonvolatile, thermally labile, and polar analytes (Zwiener and Frimmel 2004).

Several methods are available for mass analysis, including quadrupole, quadrupole ion trap, triple quadrupole, time of flight, and fourier transform ion cyclotron resonance (Niessen 2003; Schmitt-Kopplin and Frommberger 2003). Quadrupole ion trap, which applies voltages at different radiofrequencies to trap and analyze ions, has been applied to *in vivo* monitoring. For example, CLC coupled to MS is used to measure small molecule neurotransmitters, such as acetylcholine, dopamine, norepinephrine, and GABA (Zhang and Beyer 2006; Shackman et al. 2007) and the structurally similar peptides met- and leu-enkephalin (Haskins et al. 2001; Baseski et al. 2005), and to identify novel neuropeptides (Haskins et al. 2004). MS also has excellent sensitivity, and for neuropeptides the concentration detection limit is around 2 pM (Haskins et al. 2001). High sensitivity is especially important when considering that basal concentrations of neuropeptides are between 1 and 100 pM (Kennedy et al. 2002b). An example of a mass spectrogram for leu-enkephalin, along with a chromatogram from an *in vivo* measurement of met- and leu-enkephalin in the anesthetized rat, is shown in Figure 3.12B and C, respectively.

3.4.5 NEW PROBES

Two new probe types have recently been developed without a dialysis membrane. The first probe, a concentric design made from an outer stainless steel tube (410 μm diameter) and an inner fused silica capillary (50 and 150 μm inner and outer diameter, respectively), has been coupled to push–pull perfusion (Kottegoda et al. 2002). The concentric probe design has been coupled to CE–LIF for monitoring glutamate in the striatum of the anesthetized rat with a temporal resolution of 15 min. The second probe was made directly from a single fused silica capillary (90 μm diameter) and used for direct sampling of brain extracellular fluid (Kennedy et al. 2002a). The direct-sampling probe has been coupled to CE–LIF for monitoring glutamate in the striatum of the anesthetized rat as well, but with a temporal resolution of <90 s. New probe developments were made possible by the emergence of separation and detection techniques supporting low volume samples. The smaller probes improve spatial resolution and reduce tissue damage compared to conventional microdialysis probes.

3.5 RECENT ADVANCES IN BIOSENSORS

Important advances have also been made in biosensors. In addition to the attractive attributes of an electrochemical sensor, such as high temporal and spatial resolution, one important advantage of biosensors is the capability for monitoring nonelectroactive analytes. Biosensors come in many forms, but the common component is the biological recognition element that directly interacts with the analyte (Wilson and Gifford 2005). This section will focus on enzyme-linked biosensors for glutamate, where the biological recognition element is glutamate oxidase. Described below in more detail, the action of glutamate oxidase on glutamate generates an electroactive signal. However, it should be emphasized that other enzymes, such as dehydrogenases, esterases, kinases, and even ionotropic receptors, have formed the basis for biosensor designs and that a number of neurochemicals besides glutamate are amenable to detection with the biosensor approach, including acetylcholine, GABA, aspartate, glucose, lactate, and pyruvate (Hascup et al. 2007). The reader is also directed to reviews comparing glutamate measurements by microdialysis and biosensors (O'Neill et al. 1998; Kahn and Michael 2003; Watson et al. 2006).

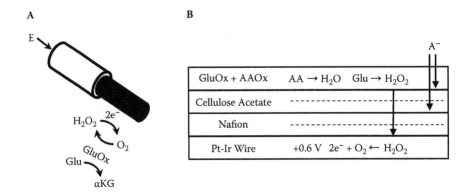

FIGURE 3.13 Enzyme-linked biosensor for glutamate. The general concept is shown in A. An early design is shown in B. Abbreviations: Glu, glutamate; GluO, glutamate oxidase; AA, ascorbic acid; AAO, ascorbic acid oxidase; αKG, α-ketoglutarate; A·, electroactive anions; Pt-Ir, platinum-iridium.

The conceptual basis for the glutamate oxidase-based biosensor is shown in Figure 3.13A. Glutamate is enzymatically acted on by glutamate oxidase, ultimately producing electroactive hydrogen peroxide. AMP is used to detect hydrogen peroxide electrochemically at the electrode surface. A series of three reactions is involved:

$$\text{glutamate} + \text{H}_2\text{O} + \text{GluOx(O)} \rightarrow \alpha\text{-ketoglutarate} + \text{NH}_3 + \text{GluOx(R)} \quad (1)$$

$$\text{GluOx(R)} + \text{O}_2 \rightarrow \text{GluOx(O)} + \text{H}_2\text{O}_2 \quad (2)$$

$$\text{H}_2\text{O}_2 \rightarrow \text{O}_2 + 2\text{H}^+ + 2\text{e}^- \quad (3)$$

Although glutamate oxidase is depicted in the oxidized (O) and reduced (R) states, a cofactor, flavin adenine dinucleotide, is actually what is reduced and oxidized in reactions 1 and 2, respectively (i.e., O = FAD and R = FADH$_2$).

In many respects, development of voltammetric electrodes and enzyme-linked biosensors shares many concerns, with chemical resolution the chief of these. The issue is not the biological recognition element, glutamate oxidase, which is exquisitely selective. Rather, amperometry offers little electrochemical information to resolve hydrogen peroxide from the plethora of easily oxidizable species in brain extracellular fluid. Taking the lead from voltammetry, negatively charged films and other enzymes have been incorporated into glutamate biosensors to address the issue of interferents.

One of the earliest glutamate biosensors was fabricated by first applying Nafion and cellulose acetate to a platinum-iridium wire (Hu et al. 1994). An enzyme layer of glutamate oxidase and ascorbic acid oxidase was then immobilized on this negatively charged film using bovine serum albumin and glutaraldehyde. The mechanism of this biosensor is shown schematically in Figure 3.13B. Negatively charged interferents are repelled by Nafion and cellulose acetate, whereas ascorbic acid is also acted on by ascorbic acid oxidase, producing the nonelectroactive water and dehydroascorbate. Most importantly, hydrogen peroxide generated by the glutamate–glutamate oxidase reaction can diffuse to the electrode for electrolysis. The sensing surface is a cylinder with a diameter of 170 μm and length of 700 μm. A temporal resolution of 1 s for glutamate is exhibited. Monoamines are also detected, but some selectivity is conferred by the slow diffusion of these positively charged neurotransmitters through the negatively charged film. Described next, subsequent glutamate biosensors have improved on this basic design.

One general concern for biosensors based on oxidases is artifacts produced by fluctuating oxygen levels, a cofactor for the enzyme. An important goal for developing glutamate biosensors is thus to reduce oxygen dependency. Interestingly, enzyme loading and miniaturization have recently been found to be beneficial in this regard (McMahon and O'Neill 2005; McMahon et al. 2006). To create a smaller probe, glutamate biosensors were fabricated on the cut end of a platinum wire, yielding a disk-shaped sensing surface with a diameter of 125 μm. Glutamate oxidase was immobilized to the surface using a polymer consisting of o-phenylenediamine and bovine serum albumin. This electrode showed high sensitivity to glutamate, but a relative lack of oxygen interference down to 5 μM, well above the 50 μM average concentration of oxygen in the brain.

To reduce size and increase selectivity, glutamate biosensors have also been constructed with carbon-fiber electrodes using "polymer wires" to transfer electrons to the sensing surface (Kulagina et al. 1999; Cui et al. 2001). The general design and mechanism for this biosensor are shown in Figure 3.14. A polymer containing an osmium-redox complex is used to immobilize three enzymes, glutamate oxidase, horseradish peroxidase and ascorbic acid oxidase, on a cylinder carbon-fiber electrode with a diameter of 10 μm and length of ~350 μm. The electrode is then coated with Nafion. The key innovation in this design is that hydrogen peroxide, formed by the glutamate–glutamate oxidase reaction, is oxidized by horseradish peroxidase, directly transferring electrons to osmium. Because a negative potential −0.1 V) is applied, the reduced osmium is subsequently oxidized, transferring electrons to the electrode. The negative potential also enhances selectivity by limiting oxidation of extracellular electroactive species. As shown in Figure 3.14B, *in vivo* glutamate measurements with this biosensor are TTX-sensitive, indicating a neuronal origin for the analyte.

One final glutamate biosensor, the so-called ceramic-based multisite electrode, is described. The original design contained four platinum recording sites, each 50 × 50 μm and spaced 200 μm apart, on a ceramic probe that tapers to a 2–5 μm tip (Burmeister et al. 2000). As shown in Figure 3.15A, newer designs have increased sensing surface area (50 × 150 μm) and reduced spacing between (50 μm edge to edge; Burmeister et al. 2002). The sensing mechanism is schematically shown in Figure 3.15B. Glutamate oxidase is immobilized on the Nafion-coated platinum surface by bovine serum albumin and glutaraldehyde. In addition to multiple recording sites at a single probe, this design also supports self-referencing (Burmeister and Gerhardt 2001). Typically used by a single probe vibrating between two fixed positions, self-referencing is a powerful technique for increasing sensitivity and characterizing analyte flux (Porterfield 2007). For self-referencing by the glutamate biosensor, one of the recording sites becomes the sentinel electrode, which does not contain glutamate oxidase (Figure 3.15B). Subtracting the sentinel signal from that

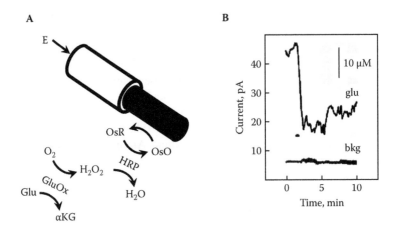

FIGURE 3.14 Enzyme-linked biosensor for glutamate based on an osmium containing redox polymer for directly transferring electrons to the platinum (Pt) electrode. Key features of the conceptual design are shown in A. The TTX-sensitivity of a response recorded in the striatum of an anesthetized rat is shown in B. TTX administration is denoted by the solid line under the top trace, labeled glu for glutamate recording. The bottom trace, labeled bkg for background, was recorded at a second electrode fabricated identically but without glutamate oxidase. Abbreviations: Glu, glutamate; GluO, glutamate oxidase; AA, ascorbic acid; AAO, ascorbic acid oxidase; αKG, α-ketoglutarate; A⁻, electroactive anions; Pt-Ir, platinum-iridium, horseradish peroxidase; OsR, reduced osmium; OsO, oxidized osmium. From Kulagina et al. (1999, p. 5093), with permission. (With permission from Kulagina NV, Shankar L, Michael AC (1999) Monitoring glutamate and ascorbate in the extracellular space of brain tissue with electrochemical microsensors. *Anal Chem* 71: 5093–5100.)

measured at the site containing glutamate oxidase removes common intereferents, such as monoamines (Figure 3.16A) and noise (Figure 3.16B), thereby increasing the quality of glutamate recording.

3.6 CONCLUDING REMARKS

The emergence of rapid sampling voltammetric techniques have clearly revolutionized the field of chemical monitoring in the last decade or so. Initially developed by a few investigators, these approaches have found their way to many other laboratories and are now used routinely for the study of neurotransmitter dynamics in anesthetized and brain slice preparations. This trend will no doubt continue with further collaboration and instrument commercialization. Because assessing neural substrates of behavior in real time is an exciting direction that is only beginning to flourish, enormous growth in these types of measurements is additionally expected. Microdialysis and biosensors must be considered in any discussion of the future of chemical monitoring. Advances in sensitivity and capabilities for multianalyte measurements, combined with new probes, bode well for the continued success of microdialysis, especially as these new approaches become more accessible to neurobiologists. The opportunity for measuring nonelectroactive neurotransmitters rapidly at a small probe makes biosensors attractive as

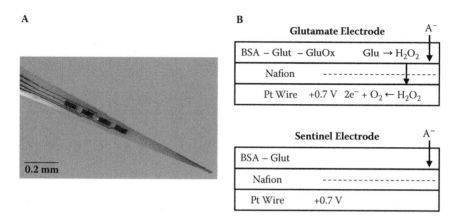

FIGURE 3.15 Ceramic-based multisite electrode. An electron micrograph of the probe is shown in A. The mechanism of glutamate sensing is shown in B. Abbreviations: Glu, glutamate; GluO, glutamate oxidase; AA, ascorbic acid; AAO, ascorbic acid oxidase; αKG, α-ketoglutarate; A⁻, electroactive anions; Pt-Ir, platinum-iridium; Pt, platinum; BSA, bovine serum albumin; Glut, glutaraldehyde. (With permission from Burmeister JJ, Pomerleau F, Palmer M, Day BK, Huettl P, Gerhardt GA (2002) Improved ceramic-based multisite microelectrode for rapid measurements of L-glutamate in the CNS. *J Neurosci Methods* 119: 163–171.)

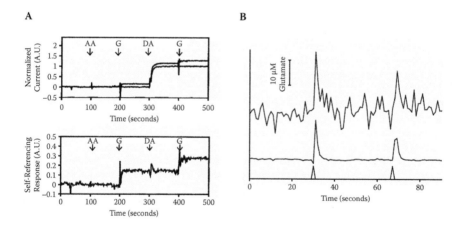

FIGURE 3.16 Self-referencing at the ceramic-based multisite electrode. The top panel in A shows the signal recorded at the sentinel and glutamate electrode. Note the response to dopamine at both electrodes, but no glutamate response at the sentinel electrode. The bottom panel in A shows the result of subtracting the sentinel signal from the glutamate electrode signal. Note that self-referencing has removed the dopamine contribution to the signal. Abbreviations: AA, ascorbic acid; G, glutamate; DA, dopamine. From Burmeister et al. (2001, p. 1037), with permission. A signal recorded in the cortex of an anesthetized rat in response to K⁺ depolarization (bottom triangles) is shown for the glutamate electrode in the top panel of B. The bottom panel of B shows subtraction of the sentinel signal. Note the improvement of signal-to-noise. (With permission from Burmeister JJ, Pomerleau F, Palmer M, Day BK, Huettl P, Gerhardt GA (2002) Improved ceramic-based multisite microelectrode for rapid measurements of L-glutamate in the CNS. *J Neurosci Methods* 119: 163–171.)

well. Just as voltammetry and microdialysis dominated the early era of chemical monitoring in the brain, these two techniques combined with biosensors will prominently factor in the study of neural substrates of addiction for many years to come.

ACKNOWLEDGMENTS

We kindly thank Drs. John Baur and Michael Heien for previewing the manuscript. This work was supported by USAMRMC 03281055, NIDA DA 021770, and NSF DBI 0138011.

REFERENCES

Adams RN (1990) In vivo electrochemical measurements in the CNS. *Prog Neurobiol* 35: 297–311.

Bakker E, Qin Y (2006) Electrochemical sensors. *Anal Chem* 78: 3965–3984.

Baseski HM, Watson CJ, Cellar NA, Shackman JG, Kennedy RT (2005) Capillary liquid chromatography with MS3 for the determination of enkephalins in microdialysis samples from the striatum of anesthetized and freely moving rats. *J Mass Spectrom* 40: 146–153.

Bath BD, Michael DJ, Trafton BJ, Joseph JD, Runnels PL, Wightman RM (2000) Subsecond adsorption and desorption of dopamine at carbon-fiber microelectrodes. *Anal Chem* 72: 5994–6002.

Baur JE, Kristensen EW, May LJ, Wiedemann DJ, Wightman RM (1988) Fast-scan voltammetry of biogenic amines. *Anal Chem* 60: 1268–1272.

Benoit-Marand M, Borrelli E, Gonon F (2001) Inhibition of dopamine release via presynaptic D2 receptors: time course and functional characteristics in vivo. *J Neurosci* 21: 9134–9141.

Benveniste H, Hansen AJ (1991) Practical aspects of using microdialysis for determination of brain interstitial concentrations. In: *Techniques in the behavioral and neural sciences: microdialysis in the neurosciences* (Robinson TE, Justice JB, eds), pp 81–102. Amsterdam: Elsevier.

Blaha CD, Phillips AG (1996) A critical assessment of electrochemical procedures applied to the measurement of dopamine and its metabolites during drug-induced and species-typical behaviours. *Behav Pharmacol* 7: 675–708.

Borland LM, Michael AC (2004) Voltammetric study of the control of striatal dopamine release by glutamate. *J Neurochem* 91: 220–229.

Borland LM, Michael AC (2007) An introduction to electrochemical methods in neuroscience. In: *Electrochemical methods for neuroscience* (Michael AC, Borland LM, eds), pp 1–15. Boca Raton, FL: CRC Press.

Boyd BW, Witowski SR, Kennedy RT (2000) Trace-level amino acid analysis by capillary liquid chromatography and application to in vivo microdialysis sampling with 10-s temporal resolution. *Anal Chem* 72: 865–871.

Budygin EA, Kilpatrick MR, Gainetdinov RR, Wightman RM (2000) Correlation between behavior and extracellular dopamine levels in rat striatum: comparison of microdialysis and fast-scan cyclic voltammetry. *Neurosci Lett* 281: 9–12.

Bunin MA, Wightman RM (1998) Quantitative evaluation of 5-hydroxytryptamine (serotonin) neuronal release and uptake: an investigation of extrasynaptic transmission. *J Neurosci* 18: 4854–4860.

Burmeister JJ, Gerhardt GA (2001) Self-referencing ceramic-based multisite microelectrodes for the detection and elimination of interferences from the measurement of L-glutamate and other analytes. *Anal Chem* 73: 1037–1042.

Burmeister JJ, Moxon K, Gerhardt GA (2000) Ceramic-based multisite microelectrodes for electrochemical recordings. *Anal Chem* 72: 187–192.

Burmeister JJ, Pomerleau F, Palmer M, Day BK, Huettl P, Gerhardt GA (2002) Improved ceramic-based multisite microelectrode for rapid measurements of L-glutamate in the CNS. *J Neurosci Methods* 119: 163–171.

Callado LF, Stamford JA (2000) Spatiotemporal interaction of alpha(2) autoreceptors and noradrenaline transporters in the rat locus coeruleus: implications for volume transmission. *J Neurochem* 74: 2350–2358.

Carelli RM, Wightman RM (2004) Functional microcircuitry in the accumbens underlying drug addiction: insights from real-time signaling during behavior. *Curr Opin Neurobiol* 14: 763–768.

Cass WA, Gerhardt GA, Mayfield RD, Curella P, Zahniser NR (1992) Differences in dopamine clearance and diffusion in rat striatum and nucleus accumbens following systemic cocaine administration. *J Neurochem* 59: 259–266.

Cass WA, Manning MW (1999) Recovery of presynaptic dopaminergic functioning in rats treated with neurotoxic doses of methamphetamine. *J Neurosci* 19: 7653–7660.

Cass WA, Zahniser NR, Flach KA, Gerhardt GA (1993) Clearance of exogenous dopamine in rat dorsal striatum and nucleus accumbens: role of metabolism and effects of locally applied uptake inhibitors. *J Neurochem* 61: 2269–2278.

Cheer JF, Aragona BJ, Heien ML, Seipel AT, Carelli RM, Wightman RM (2007a) Coordinated accumbal dopamine release and neural activity drive goal-directed behavior. *Neuron* 54: 237–244.

Cheer JF, Heien ML, Garris PA, Carelli RM, Wightman RM (2005) Simultaneous electrochemical and single-unit recordings in the nucleus accumbens reveal GABA-mediated responses: implications for intracranial self-stimulation. *Proc Natl Acad Sci U S A*.

Cheer JF, Wassum KM, Heien ML, Phillips PE, Wightman RM (2004) Cannabinoids enhance subsecond dopamine release in the nucleus accumbens of awake rats. *J Neurosci* 24: 4393–4400.

Cheer JF, Wassum KM, Sombers LA, Heien ML, Ariansen JL, Aragona BJ, Phillips PE, Wightman RM (2007b) Phasic dopamine release evoked by abused substances requires cannabinoid receptor activation. *J Neurosci* 27: 791–795.

Chen KC, Budygin EA (2007) Extracting the basal extracellular dopamine concentrations from the evoked responses: re-analysis of the dopamine kinetics. *J Neurosci Methods* 164: 27–42.

Chen Z, Wu J, Baker GB, Parent M, Dovichi NJ (2001) Application of capillary electrophoresis with laser-induced fluorescence detection to the determination of biogenic amines and amino acids in brain microdialysate and homogenate samples. *J Chromatogr A* 914: 293–298.

Chergui K, Suaud-Chagny MF, Gonon F (1994) Nonlinear relationship between impulse flow, dopamine release and dopamine elimination in the rat brain in vivo. *Neuroscience* 62: 641–645.

Chiu TC, Lin YW, Huang YF, Chang HT (2006) Analysis of biologically active amines by CE. *Electrophoresis* 27: 4792–4807.

Chow RH, von Ruden L, Neher E (1992) Delay in vesicle fusion revealed by electrochemical monitoring of single secretory events in adrenal chromaffin cells. *Nature* 356: 60–63.

Cragg S, Rice ME, Greenfield SA (1997) Heterogeneity of electrically evoked dopamine release and reuptake in substantia nigra, ventral tegmental area, and striatum. *J Neurophysiol* 77: 863–873.

Cui J, Kulagina NV, Michael AC (2001) Pharmacological evidence for the selectivity of in vivo signals obtained with enzyme-based electrochemical sensors. *J Neurosci Methods* 104: 183–189.

Davidson C, Stamford JA (1995) Evidence that 5-hydroxytryptamine release in rat dorsal raphe nucleus is controlled by 5-HT1A, 5-HT1B and 5-HT1D autoreceptors. *Br J Pharmacol* 114: 1107–1109.

Daws LC, Toney GM (2007) High-speed chronoamperometry to study kinetics and mechanisms for serotonin clearance in vivo. In: *Electrochemical methods for neuroscience* (Michael AC, Borland LM, eds), pp 63–81. Boca Raton, FL: CRC Press.

Dickinson SD, Sabeti J, Larson GA, Giardina K, Rubinstein M, Kelly MA, Grandy DK, Low MJ, Gerhardt GA, Zahniser NR (1999) Dopamine D2 receptor-deficient mice exhibit decreased dopamine transporter function but no changes in dopamine release in dorsal striatum. *J Neurochem* 72: 148–156.

Dommett E, Coizet V, Blaha CD, Martindale J, Lefebvre V, Walton N, Mayhew JE, Overton PG, Redgrave P (2005) How visual stimuli activate dopaminergic neurons at short latency. *Science* 307: 1476–1479.

Dugast C, Suaud-Chagny MF, Gonon F (1994) Continuous in vivo monitoring of evoked dopamine release in the rat nucleus accumbens by amperometry. *Neuroscience* 62: 647–654.

Falkenburger BH, Barstow KL, Mintz IM (2001) Dendrodendritic inhibition through reversal of dopamine transport. *Science* 293: 2465–2470.

Forster GL, Blaha CD (2003) Pedunculopontine tegmental stimulation evokes striatal dopamine efflux by activation of acetylcholine and glutamate receptors in the midbrain and pons of the rat. *Eur J Neurosci* 17: 751–762.

Garcia-Campana AM, Taverna M, Fabre H (2007) LIF detection of peptides and proteins in CE. *Electrophoresis* 28: 208–232.

Garris PA, Budygin EA, Phillips PE, Venton BJ, Robinson DL, Bergstrom BP, Rebec GV, Wightman RM (2003) A role for presynaptic mechanisms in the actions of nomifensine and haloperidol. *Neuroscience* 118: 819–829.

Garris PA, Christensen JR, Rebec GV, Wightman RM (1997) Real-time measurement of electrically evoked extracellular dopamine in the striatum of freely moving rats. *J Neurochem* 68: 152–161.

Garris PA, Kilpatrick M, Bunin MA, Michael D, Walker QD, Wightman RM (1999) Dissociation of dopamine release in the nucleus accumbens from intracranial self-stimulation. *Nature* 398: 67–69.

Garris PA, Wightman RM (1994) Different kinetics govern dopaminergic transmission in the amygdala, prefrontal cortex, and striatum: an in vivo voltammetric study. *J Neurosci* 14: 442–450.

Gerhardt GA (1995) Rapid chronocoulometric measurements of norepinephrine overflow and clearance in CNS tissues. In: *Neuromethods: voltammetric methods in brain systems* (Boulton A, Baker G, Adams RN, eds), pp 117–151. Totowa, NJ: Humana Press.

Gerhardt GA, Cass WA, Henson M, Zhang Z, Ovadia A, Hoffer BJ, Gash DM (1995) Age-related changes in potassium-evoked overflow of dopamine in the striatum of the rhesus monkey. *Neurobiol Aging* 16: 939–946.

Gerhardt GA, Hoffman AF (2001) Effects of recording media composition on the responses of Nafion-coated carbon fiber microelectrodes measured using high-speed chronoamperometry. *J Neurosci Methods* 109: 13–21.

Gerhardt GA, Ksir C, Pivik C, Dickinson SD, Sabeti J, Zahniser NR (1999) Methodology for coupling local application of dopamine and other chemicals with rapid in vivo electrochemical recordings in freely moving rats. *J Neurosci Methods* 87: 67–76.

Gerhardt GA, Palmer MR (1987) Characterization of the techniques of pressure ejection and microiontophoresis using in vivo electrochemistry. *J Neurosci Methods* 22: 147–159.

Ghasemzadeh MB, Capella P, Mitchell K, Adams RN (1993) Real-time monitoring of electrically stimulated norepinephrine release in rat thalamus: I. Resolution of transmitter and metabolite signal components. *J Neurochem* 60: 442–448.

Gonon F (1995) Monitoring dopamine and noradrenaline release in central and peripheral nervous systems with treated and untreated carbon-fiber electrodes. In: *Neuromethods: voltammetric methods in brain systems* (Boulton A, Baker G, Adams RN, eds), pp 153–177. Totowa, NJ: Humana Press, Inc.

Gonon F (1997) Prolonged and extrasynaptic excitatory action of dopamine mediated by D1 receptors in the rat striatum in vivo. *J Neurosci* 17: 5972–5978.

Gonon F, Sundstrom L (1996) Excitatory effects of dopamine released by impulse flow in the rat nucleus accumbens in vivo. *Neuroscience* 75: 13–18.

Gulley JM, Larson GA, Zahniser NR (2007) *Using high-speed chronoamperometry with local dopamine application to assess dopamien transporter function.* Boca Raton, FL: CRC Press.

Hascup KN, Rutherford EC, Quintero JE, Day BK, Nickell JR, Pomerleau F, Huettl P, Burmeister JJ, Gerhardt GA (2007) Second-by-second measures of L-glutamate and other neurotransmitters using enzyme-based microelectrode arrays. In: *Electrochemical methods for neuroscience* (Michael AC, Borland LM, eds), pp 407–450. Boca Raton: CRC Press.

Haskins WE, Wang Z, Watson CJ, Rostand RR, Witowski SR, Powell DH, Kennedy RT (2001) Capillary LC-MS2 at the attomole level for monitoring and discovering endogenous peptides in microdialysis samples collected in vivo. *Anal Chem* 73: 5005–5014.

Haskins WE, Watson CJ, Cellar NA, Powell DH, Kennedy RT (2004) Discovery and neurochemical screening of peptides in brain extracellular fluid by chemical analysis of in vivo microdialysis samples. *Anal Chem* 76: 5523–5533.

Heien ML, Johnson MA, Wightman RM (2004) Resolving neurotransmitters detected by fast-scan cyclic voltammetry. *Anal Chem* 76: 5697–5704.

Heien ML, Khan AS, Ariansen JL, Cheer JF, Phillips PE, Wassum KM, Wightman RM (2005) Real-time measurement of dopamine fluctuations after cocaine in the brain of behaving rats. *Proc Natl Acad Sci U S A* 102: 10023–10028.

Heien ML, Phillips PE, Stuber GD, Seipel AT, Wightman RM (2003) Overoxidation of carbon-fiber microelectrodes enhances dopamine adsorption and increases sensitivity. *Analyst* 128: 1413–1419.

Hernandez-Borges J, Neususs C, Cifuentes A, Pelzing M (2004) On-line capillary electrophoresis-mass spectrometry for the analysis of biomolecules. *Electrophoresis* 25: 2257–2281.

Hoffman AF, Lupica CR, Gerhardt GA (1998) Dopamine transporter activity in the substantia nigra and striatum assessed by high-speed chronoamperometric recordings in brain slices. *J Pharmacol Exp Ther* 287: 487–496.

Hu Y, Mitchell KM, Albahadily FN, Michaelis EK, Wilson GS (1994) Direct measurement of glutamate release in the brain using a dual enzyme-based electrochemical sensor. *Brain Res* 659: 117–125.

Hyman SE (2005) Addiction: a disease of learning and memory. *Am J Psychiatry* 162: 1414–1422.

Issaq HJ (1999) Capillary electrophoresis of natural products—II. *Electrophoresis* 20: 3190–3202.

Jones SR, Gainetdinov RR, Wightman RM, Caron MG (1998) Mechanisms of amphetamine action revealed in mice lacking the dopamine transporter. *J Neurosci* 18: 1979–1986.

Jones SR, Garris PA, Kilts CD, Wightman RM (1995) Comparison of dopamine uptake in the basolateral amygdaloid nucleus, caudate-putamen, and nucleus accumbens of the rat. *J Neurochem* 64: 2581–2589.

Jones SR, Joseph JD, Barak LS, Caron MG, Wightman RM (1999) Dopamine neuronal transport kinetics and effects of amphetamine. *J Neurochem* 73: 2406–2414.

Jones SR, Mickelson GE, Collins LB, Kawagoe KT, Wightman RM (1994) Interference by pH and Ca2+ ions during measurements of catecholamine release in slices of rat amygdala with fast-scan cyclic voltammetry. *J Neurosci Methods* 52: 1–10.

Joseph MH (1996) Can voltammetric electrodes measure brain extracellular dopamine in the behaving animal? Commentary on Blaha and Phillips', "A critical assessment of electrochemical procedures applied to the measurement of dopamine and its metabolites during drug-induced and species-typical behaviours." *Behav Pharmacol* 7: 709–713.

Justice JB (1987) Introduction to in vivo voltammetry. In: *Voltammetry in the neurosciences* (Justice JB, ed), pp 3–101. Clifton, NJ: Humana Press.

Justice JBJ (1993) Quantitative microdialysis of neurotransmitters. *J Neurosci Methods* 48: 263–276.

Kahn AS, Michael AC (2003) Invasive consequences of using micro-electrodes and microdialysis probes in the brain. *Trends Anal Chem* 22: 503–508.

Kalivas PW, Volkow ND (2005) The neural basis of addiction: a pathology of motivation and choice. *Am J Psychiatry* 162: 1403–1413.

Kawagoe KT, Garris PA, Wiedemann DJ, Wightman RM (1992) Regulation of transient dopamine concentration gradients in the microenvironment surrounding nerve terminals in the rat striatum. *Neuroscience* 51: 55–64.

Kawagoe KT, Wightman RM (1994) Characterization of amperometry for in vivo measurement of dopamine dynamics in the rat brain. *Talanta* 41: 865–874.

Kawagoe KT, Zimmerman JB, Wightman RM (1993) Principles of voltammetry and microelectrode surface states. *J Neurosci Methods* 48: 225–240.

Kennedy RT, Thompson JE, Vickroy TW (2002a) In vivo monitoring of amino acids by direct sampling of brain extracellular fluid at ultralow flow rates and capillary electrophoresis. *J Neurosci Methods* 114: 39–49.

Kennedy RT, Watson CJ, Haskins WE, Powell DH, Strecker RE (2002b) In vivo neurochemical monitoring by microdialysis and capillary separations. *Curr Opin Chem Biol* 6: 659–665.

Kilpatrick MR, Rooney MB, Michael DJ, Wightman RM (2000) Extracellular dopamine dynamics in rat caudate-putamen during experimenter-delivered and intracranial self-stimulation. *Neuroscience* 96: 697–706.

Kiyatkin EA, Kiyatkin DE, Rebec GV (2000) Phasic inhibition of dopamine uptake in nucleus accumbens induced by intravenous cocaine in freely behaving rats. *Neuroscience* 98: 729–741.

Kiyatkin EA, Wise RA, Gratton A (1993) Drug- and behavior-associated changes in dopamine-related electrochemical signals during intravenous heroin self-administration in rats. *Synapse* 14: 60–72.

Kottegoda S, Shaik I, Shippy SA (2002) Demonstration of low flow push–pull perfusion. *J Neurosci Methods* 121: 93–101.

Kristensen EW, Wilson RM, Wightman RM (1986) Dispersion in flow injection analysis measured with microvoltammetric electrodes. *Anal Chem* 58: 986–988.

Kulagina NV, Shankar L, Michael AC (1999) Monitoring glutamate and ascorbate in the extracellular space of brain tissue with electrochemical microsensors. *Anal Chem* 71: 5093–5100.

Kulagina NV, Zigmond MJ, Michael AC (2001) Glutamate regulates the spontaneous and evoked release of dopamine in the rat striatum. *Neuroscience* 102: 121–128.

Lada MW, Kennedy RT (1996) Quantitative in vivo monitoring of primary amines in rat caudate nucleus using microdialysis coupled by a flow-gated interface to capillary electrophoresis with laser-induced fluorescence detection. *Anal Chem* 68: 2790–2797.

Lee KH, Blaha CD, Harris BT, Cooper S, Hitti FL, Leiter JC, Roberts DW, Kim U (2006) Dopamine efflux in the rat striatum evoked by electrical stimulation of the subthalamic nucleus: potential mechanism of action in Parkinson's disease. *Eur J Neurosci* 23: 1005–1014.

Marsden CA, Joseph MH, Kruk ZL, Maidment NT, O'Neill RD, Schenk JO, Stamford JA (1988) In vivo voltammetry—present electrodes and methods. *Neuroscience* 25: 389–400.

McMahon CP, O'Neill RD (2005) Polymer-enzyme composite biosensor with high glutamate sensitivity and low oxygen dependence. *Anal Chem* 77: 1196–1199.

McMahon CP, Rocchitta G, Serra PA, Kirwan SM, Lowry JP, O'Neill RD (2006) Control of the oxygen dependence of an implantable polymer/enzyme composite biosensor for glutamate. *Anal Chem* 78: 2352–2359.

Michael AC, Borland LM, Mitala JJ, Jr., Willoughby BM, Motzko CM (2005) Theory for the impact of basal turnover on dopamine clearance kinetics in the rat striatum after medial forebrain bundle stimulation and pressure ejection. *J Neurochem* 94: 1202–1211.

Millar J, Armstrong-James M, Kruk ZL (1981) Polarographic assay of iontophoretically applied dopamine and low-noise unit recording using a multibarrel carbon fibre microelectrode. *Brain Res* 205: 419–424.

Millar J, Stamford JA, Kruk ZL, Wightman RM (1985) Electrochemical, pharmacological and electrophysiological evidence of rapid dopamine release and removal in the rat caudate nucleus following electrical stimulation of the median forebrain bundle. *Eur J Pharmacol* 109: 341–348.

Millar J, Williams GV (1988) Ultra-low noise silver-plated carbon fibre microelectrodes. *J Neurosci Methods* 25: 59–62.

Mitchell K, Oke AF, Adams RN (1994) In vivo dynamics of norepinephrine release-reuptake in multiple terminal field regions of rat brain. *J Neurochem* 63: 917–926.

Moini M (2002) Capillary electrophoresis mass spectrometry and its application to the analysis of biological mixtures. *Anal BioAnal Chem* 373: 466–480.

Nicholson C (1985) Diffusion from an injected volume of a substance in brain tissue with arbitrary volume fraction and tortuosity. *Brain Res* 333: 325–329.

Niessen WM (2003) Progress in liquid chromatography-mass spectrometry instrumentation and its impact on high-throughput screening. *J Chromatogr A* 1000: 413–436.

O'Neill RD, Lowry JP, Mas M (1998) Monitoring brain chemistry in vivo: voltammetric techniques, sensors, and behavioral applications. *Crit Rev Neurobiol* 12: 69–127.

Olds J, Milner PM (1954) Positive reinforcement produced by electrical stimulation of septal area and other regions of the rat brain. *J Comp Physiol Psychol* 47: 419–427.

Overton PG, Clark D (1997) Burst firing in midbrain dopaminergic neurons. *Brain Res Brain Res Rev* 25: 312–334.

Palij P, Bull DR, Sheehan MJ, Millar J, Stamford J, Kruk ZL, Humphrey PP (1990) Presynaptic regulation of dopamine release in corpus striatum monitored in vitro in real time by fast cyclic voltammetry. *Brain Res* 509: 172–174.

Palij P, Stamford JA (1992) Real-time monitoring of endogenous noradrenaline release in rat brain slices using fast cyclic voltammetry: 1. Characterisation of evoked noradrenaline efflux and uptake from nerve terminals in the bed nucleus of stria terminalis, pars ventralis. *Brain Res* 587: 137–146.

Parrot S, Sauvinet V, Riban V, Depaulis A, Renaud B, Denoroy L (2004) High temporal resolution for in vivo monitoring of neurotransmitters in awake epileptic cat using brain microdialysis and capillary electrophoresis with laser-induced fluorescence detection. *J Neurosci* 140: 29–38.

Phillips PE, Robinson DL, Stuber GD, Carelli RM, Wightman RM (2003) Real-time measurements of phasic changes in extracellular dopamine concentration in freely moving rats by fast-scan cyclic voltammetry. *Methods Mol Med* 79: 443–464.

Phillips PEM, Wightman RM (2003) Critical guidelines for validation of the selectivity of in-vivo chemical microsensors. *Trends Anal Chem* 22: 509–514.

Porterfield DM (2007) Measuring metabolism and biophysical flux in the tissue, cellular and sub-cellular domains: recent developments in self-referencing amperometry for physiological sensing. *Biosens Bioelectron* 22: 1186–1196.

Powell PR, Ewing AG (2005) Recent advances in the application of capillary electrophoresis to neuroscience. *Anal BioAnal Chem* 382: 581–591.

Rebec GV, Christensen JR, Guerra C, Bardo MT (1997) Regional and temporal differences in real-time dopamine efflux in the nucleus accumbens during free-choice novelty. *Brain Res* 776: 61–67.

Rebec GV, Witowski SR, Sandstrom MI, Rostand RD, Kennedy RT (2005) Extracellular ascorbate modulates cortically evoked glutamate dynamics in rat striatum. *Neurosci Lett* 378: 166–170.

Rice ME, Cragg SJ (2004) Nicotine amplifies reward-related dopamine signals in striatum. *Nat Neurosci* 7: 583–584.

Rice ME, Cragg SJ, Greenfield SA (1997) Characteristics of electrically evoked somatodendritic dopamine release in substantia nigra and ventral tegmental area in vitro. *J Neurophysiol* 77: 853–862.

Rice ME, Nicholson C (1995) Diffusion and ion shifts in the brain extracellular microenvironment and their relevance for voltammetric measurements. In: *Neuromethods: voltammetric methods in brain systems* (Boulton A, Baker G, Adams RN, eds), pp 27–81. Totowa NJ: Humana Press.

Robinson DL, Heien ML, Wightman RM (2002) Frequency of dopamine concentration transients increases in dorsal and ventral striatum of male rats during introduction of conspecifics. *J Neurosci* 22: 10477–10486.

Robinson DL, Phillips PE, Budygin EA, Trafton BJ, Garris PA, Wightman RM (2001) Subsecond changes in accumbal dopamine during sexual behavior in male rats. *Neuroreport* 12: 2549–2552.

Robinson DL, Wightman RM (2004) Nomifensine amplifies subsecond dopamine signals in the ventral striatum of freely-moving rats. *J Neurochem* 90: 894–903.

Roitman MF, Stuber GD, Phillips PE, Wightman RM, Carelli RM (2004) Dopamine operates as a subsecond modulator of food seeking. *J Neurosci* 24: 1265–1271.

Sabeti J, Gerhardt GA, Zahniser NR (2002) Acute cocaine differentially alters accumbens and striatal dopamine clearance in low and high cocaine locomotor responders: behavioral and electrochemical recordings in freely moving rats. *J Pharmacol Exp Ther* 302: 1201–1211.

Sabeti J, Gerhardt GA, Zahniser NR (2003a) Chloral hydrate and ethanol, but not urethane, alter the clearance of exogenous dopamine recorded by chronoamperometry in striatum of unrestrained rats. *Neurosci Lett* 343: 9–12.

Sabeti J, Gerhardt GA, Zahniser NR (2003b) Individual differences in cocaine-induced locomotor sensitization in low and high cocaine locomotor-responding rats are associated with differential inhibition of dopamine clearance in nucleus accumbens. *J Pharmacol Exp Ther* 305: 180–190.

Schmitt-Kopplin P, Frommberger M (2003) Capillary electrophoresis-mass spectrometry: 15 years of developments and applications. *Electrophoresis* 24: 3837–3867.

Schmitz Y, Lee CJ, Schmauss C, Gonon F, Sulzer D (2001) Amphetamine distorts stimulation-dependent dopamine overflow: effects on D2 autoreceptors, transporters, and synaptic vesicle stores. *J Neurosci* 21: 5916–5924.

Schonfuss D, Reum T, Olshausen P, Fischer T, Morgenstern R (2001) Modelling constant potential amperometry for investigations of dopaminergic neurotransmission kinetics in vivo. *J Neurosci Methods* 112: 163–172.

Schultz W (1998) Predictive reward signal of dopamine neurons. *J Neurophysiol* 80: 1–27.

Shackman HM, Shou M, Cellar NA, Watson CJ, Kennedy RT (2007) Microdialysis coupled on-line to capillary liquid chromatography with tandem mass spectrometry for monitoring acetylcholine in vivo. *J Neurosci Methods* 159: 86–92.

Shen H, Lada MW, Kennedy RT (1997) Monitoring of met-enkephalin in vivo with 5-min temporal resolution using microdialysis sampling and capillary liquid chromatography with electrochemical detection. *J Chromatogr B Biomed Sci Appl* 704: 43–52.

Shou M, Ferrario CR, Schultz KN, Robinson TE, Kennedy RT (2006) Monitoring dopamine in vivo by microdialysis sampling and on-line CE-laser-induced fluorescence. *Anal Chem* 78: 6717–6725.

Stamford JA (1989) In vivo voltammetry—prospects for the next decade. *Trends Neurosci* 12: 407–412.

Stamford JA, Palij P, Davidson C, Jorm CM, Millar J (1993) Simultaneous real-time electrochemical and electrophysiological recording in brain slices with a single carbon-fibre microelectrode. *J Neurosci Methods* 50: 279–290.

Stuber GD, Roitman MF, Phillips PE, Carelli RM, Wightman RM (2005a) Rapid dopamine signaling in the nucleus accumbens during contingent and noncontingent cocaine administration. *Neuropsychopharmacology* 30: 853–863.

Stuber GD, Wightman RM, Carelli RM (2005b) Extinction of cocaine self-administration reveals functionally and temporally distinct dopaminergic signals in the nucleus accumbens. *Neuron* 46: 661–669.

Suaud-Chagny MF, Chergui K, Chouvet G, Gonon F (1992) Relationship between dopamine release in the rat nucleus accumbens and the discharge activity of dopaminergic neurons during local in vivo application of amino acids in the ventral tegmental area. *Neuroscience* 49: 63–72.

Suaud-Chagny MF, Dugast C, Chergui K, Msghina M, Gonon F (1995) Uptake of dopamine released by impulse flow in the rat mesolimbic and striatal systems in vivo. *J Neurochem* 65: 2603–2611.

Takeuchi T (2005) Development of capillary liquid chromatography. *Chromatography* 26.

Thurman EM, Ferrer I, Barcelo D (2001) Choosing between atmospheric pressure chemical ionization and electrospray ionization interfaces for the HPLC/MS analysis of pesticides. *Anal Chem* 73: 5441–5449.

Tucci S, Rada P, Sepulveda MJ, Hernandez L (1997) Glutamate measured by 6-s resolution brain microdialysis: capillary electrophoretic and laser-induced fluorescence detection application. *J Chromatogr B Biomed Sci Appl* 694: 343–349.

Ungerstedt U (1991) Microdialysis—principles and applications for studies in animals and man. *J Intern Med* 230: 365–373.

Venton BJ, Michael DJ, Wightman RM (2003a) Correlation of local changes in extracellular oxygen and pH that accompany dopaminergic terminal activity in the rat caudate-putamen. *J Neurochem* 84: 373–381.

Venton BJ, Robinson TE, Kennedy RT (2006a) Transient changes in nucleus accumbens amino acid concentrations correlate with individual responsivity to the predator fox odor 2,5-dihydro-2,4,5-trimethylthiazoline. *J Neurochem* 96: 236–246.

Venton BJ, Robinson TE, Kennedy RT, Maren S (2006b) Dynamic amino acid increases in the basolateral amygdala during acquisition and expression of conditioned fear. *Eur J Neurosci* 23: 3391–3398.

Venton BJ, Troyer KP, Wightman RM (2002) Response times of carbon fiber microelectrodes to dynamic changes in catecholamine concentration. *Anal Chem* 74: 539–546.

Venton BJ, Zhang H, Garris PA, Phillips PE, Sulzer D, Wightman RM (2003b) Real-time decoding of dopamine concentration changes in the caudate-putamen during tonic and phasic firing. *J Neurochem* 87: 1284–1295.

Walker QD, Rooney MB, Wightman RM, Kuhn CM (2000) Dopamine release and uptake are greater in female than male rat striatum as measured by fast cyclic voltammetry. *Neuroscience* 95: 1061–1070.

Watson CJ, Venton BJ, Kennedy RT (2006) In vivo measurements of neurotransmitters by microdialysis sampling. *Anal Chem* 78: 1391–1399.

Westerink BH (1995) Brain microdialysis and its application for the study of animal behaviour. *Behav Brain Res* 70: 103–124.

Wightman RM, Amatore C, Engstrom RC, Hale PD, Kristensen EW, Kuhr WG, May LJ (1988) Real-time characterization of dopamine overflow and uptake in the rat striatum. *Neuroscience* 25: 513–523.

Wightman RM, Brown DS, Kuhr WG, Wilson RL (1987) Molecular specificity of in vivo electrochemical measurements. In: *Voltammetry in the neurosciences* (Justice JB, ed), pp 103–138. Clifton, NJ: Human Press.

Wightman RM, Jankowski JA, Kennedy RT, Kawagoe KT, Schroeder TJ, Leszczyszyn DJ, Near JA, Diliberto EJJ, Viveros OH (1991) Temporally resolved catecholamine spikes correspond to single vesicle release from individual chromaffin cells. *Proc Natl Acad Sci U S A* 88: 10754–10758.

Williams GV, Millar J (1990a) Concentration-dependent actions of stimulated dopamine release on neuronal activity in rat striatum. *Neuroscience* 39: 1–16.

Williams GV, Millar J (1990b) Differential actions of endogenous and iontophoretic dopamine in rat striatum. *Eur J Neurosci* 2: 658–661.

Wilson GS, Gifford R (2005) Biosensors for real-time in vivo measurements. *Biosens Bioelectron* 20: 2388–2403.

Wise RA (2005) Forebrain substrates of reward and motivation. *J Comp Neurol* 493: 115–121.

Wu Q, Reith ME, Kuhar MJ, Carroll FI, Garris PA (2001a) Preferential increases in nucleus accumbens dopamine after systemic cocaine administration are caused by unique characteristics of dopamine neurotransmission. *J Neurosci* 21: 6338–6347.

Wu Q, Reith ME, Walker QD, Kuhn CM, Carroll FI, Garris PA (2002) Concurrent autoreceptor-mediated control of dopamine release and uptake during neurotransmission: an in vivo voltammetric study. *J Neurosci* 22: 6272–6281.

Wu Q, Reith ME, Wightman RM, Kawagoe KT, Garris PA (2001b) Determination of release and uptake parameters from electrically evoked dopamine dynamics measured by real-time voltammetry. *J Neurosci Methods* 112: 119–133.

Yavich L, Tiihonen J (2000a) In vivo voltammetry with removable carbon fibre electrodes in freely-moving mice: dopamine release during intracranial self-stimulation. *J Neurosci Methods* 104: 55–63.

Yavich L, Tiihonen J (2000b) Patterns of dopamine overflow in mouse nucleus accumbens during intracranial self-stimulation. *Neurosci Lett* 293: 41–44.

Young SD, Michael AC (1993) Voltammetry of extracellular dopamine in rat striatum during ICSS-like electrical stimulation of the medial forebrain bundle. *Brain Res* 600: 305–307.

Zahniser NR, Dickinson SD, Gerhardt GA (1998) High-speed chronoamperometric electrochemical measurements of dopamine clearance. *Methods Enzymol* 296: 708–719.

Zahniser NR, Larson GA, Gerhardt GA (1999) In vivo dopamine clearance rate in rat striatum: regulation by extracellular dopamine concentration and dopamine transporter inhibitors. *J Pharmacol Exp Ther* 289: 266–277.

Zhang H, Sulzer D (2004) Frequency-dependent modulation of dopamine release by nicotine. *Nat Neurosci* 7: 581–582.

Zhang MY, Beyer CE (2006) Measurement of neurotransmitters from extracellular fluid in brain by in vivo microdialysis and chromatography-mass spectrometry. *J Pharm Biomed Anal* 40: 492–499.

Zwiener C, Frimmel FH (2004) LC-MS analysis in the aquatic environment and in water treatment—a critical review. Part I: Instrumentation and general aspects of analysis and detection. *Anal BioAnal Chem* 378: 851–861.

4 Alcohol Craving and Relapse Prediction
Imaging Studies

Andreas Heinz, Anne Beck, Jan Mir, Sabine M. Grüsser, Anthony A. Grace, and Jana Wrase

CONTENTS

4.1 INTRODUCTION AND SCOPE

After detoxification, patients suffering from a drug addiction show high relapse risks. For example, among alcoholics up to 85% of all patients relapse, independent of whether they have been treated as inpatients until complete remission of physical withdrawal symptoms (Boothby and Doering 2005). Current brain-imaging studies have tried to identify the neuronal correlates of learning processes involved in relapse such as behavioral mechanisms including classical Pavlovian and instrumental conditioning. The goal of such studies is to provide insight into the neurobiology of drug addiction as well as to provide new options for specific behavioral intervention or pharmacological modification of alcohol craving and the risk of relapse. In this chapter, we will discuss the theoretical background and the results of neuroimaging studies that identified alterations and relevant neurotransmitter systems that are associated with cue-induced brain activation and alcohol craving, and their impact on neuronal networks that are activated by drug-specific cues.

4.2 METHODICAL APPROACHES TO STUDY NEURONAL SYSTEMS RELEVANT FOR ALCOHOL ADDICTION

Blood oxygen level dependent (BOLD) functional magnetic resonance imaging (fMRI) studies allow us to assess brain activity by measuring the ratio of oxygenated and deoxygenated blood in humans. Functional MRI measures the hemodynamic response related to neuronal activity in the brain and is one of the most prominent neuroimaging techniques because of its noninvasiveness, lack of radiation exposure, and relatively wide availability. Changes in blood flow and blood oxygenation in the brain are closely linked to neuronal activity (Kwong et al. 1992; Logothetis 2002; Ogawa et al. 1990). Neuronal activity causes an immediate need for energy; that is, oxygen. The local response to this oxygen utilization is an increase in blood flow to regions of increased neuronal activity, occurring after a delay of approximately 5 seconds. This hemodynamic response rises to a peak before falling back to baseline and typically to a slight undershoot. This leads to local changes in the relative concentration of oxyhemoglobin and deoxyhemoglobin, and changes in local cerebral blood volume and local cerebral blood flow. The difference in magnetic susceptibility (degree of magnetization) between oxyhemoglobin and deoxyhemoglobin leads to magnetic signal variation, which can be detected using an MRI scanner.

In the neurosciences, subjects are commonly exposed to alternating presentations of, for example, affective stimuli (neutral, positive, and negative words) and a fixation (baseline) condition (crosshair). This design enables investigators to examine the main effects of brain activity related to different stimuli relative to a neutral condition (e.g., negative word vs. fixation). In this context the nature of the stimuli that are used, and their explicit versus implicit emotional valence is a critical issue, since all data reflect relative differences and there is no absolute baseline. A passive resting or fixation cross baseline condition under the conditions of fMRI scanning appears not be emotionally neutral, because it involves unpredictable mental states (Gusnard and Raichle 2001; Stark and Squire 2001) and environmental confinement and noise, and the relatively unconstrained experience of the subjects during the presentation of baseline conditions may be more uncertain and ambiguous than familiar unambiguous stimuli (e.g., neutral words). For example, a study by our group showed that unconstrained processing of an affectively undefined symbol such as a fixation cross in the potentially anxiogenic (narrow, dark, and loud) environment of an MRI scanner elicited increased amygdala activation in serotonin transporter (5HTT) reduced function s (short) carriers versus ll (long/long) homozygotes compared with relatively more visual processing of affectively neutral stimuli (Heinz et al. 2007b).

fMRI offers several advantages including high spatial resolution and noninvasiveness as well as an economic utilization. A disadvantage of this method is that effects of different stimuli on brain activation (particularly neuronal afferent input) are only indirectly reflected in the BOLD response (Logothetis 2002). Furthermore, there is rather poor temporal resolution (e.g., versus EEG), and specific neurotransmitter systems can only be assessed by measuring effects of systemically applied agonists and antagonists.

In contrast to the measurement of cerebral blood flow with fMRI, positron emission tomography (PET) as well as single photon emission computed tomography (SPECT) allow for measurement of an absolute baseline of activation. Furthermore, they can quantify neuroreceptor and transporter availabilities (with the potential confound of endogenous neurotransmitter concentration competing for binding of receptors/transporters with radioligands [Heinz et al. 2004b; Kumakura et al. 2007; Laruelle 2000]). An example is the radioligand [^{123}I]beta-CIT, which is widely used with SPECT to measure the availability of central dopamine and serotonin transporters. Such monoamine transporters regulate the reuptake of released neurotransmitters. In a study of our group, neurotransmitter concentrations measured *in vivo* with microdialysis were negatively correlated with the availability of monoamine transporters measured with [^{123}I]beta-CIT and SPECT (Heinz et al. 1999). However, this negative correlation could also be caused by a competition between the endogenous neurotransmitter and the radioligand for transporter binding sites (Fisher et al. 1995) (see Figure 4.1). This finding suggests that radioligand binding reflects the availability of D2 receptors that are currently not occupied by endogenous dopamine rather than the absolute density of D2 receptors.

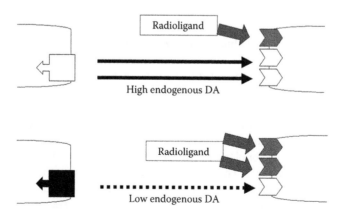

FIGURE 4.1 Blockade of dopamine D2 receptors by endogenous dopamine (top: high-endogenous dopamine and low-radioligand binding; bottom: low-endogenous dopamine and high-radioligand binding).

Disadvantages are the application of a radioactive contrast agent, which exposes the patient to gamma radiation, and the rather short half-life period of the tracer, as well as the high expenses and the complex measuring procedure.

Another methodological approach is the combination of both methods: the correlation of neuronal activation measured with fMRI and functions of neurotransmitter system measured with PET or SPECT (e.g., dopamine receptor D2 availability or dopamine synthesis rate). This is an elegant way to link certain brain functions with certain neurotransmitter systems, like linking dopamine with the brain network that controls reward, pleasure, and motivation.

Furthermore, human studies often use the startle response as a neurophysiological indicator of appetitive or aversive reactions toward visual stimuli. In alcohol-dependent patients, it has been observed that alcohol cues often elicit a physiological response similar to appetitive cues, which is not necessarily reflected in a conscious feeling of attraction or pleasure (Heinz et al. 2003). Indeed, conditioned reactions can be assessed on multiple levels and these levels of reactions differ conceptually and can be influenced by the conditioned stimulus with different intensity (Carter and Tiffany 1999).

4.3 NEUROTRANSMITTER SYSTEMS IMPLICATED IN CRAVING FOR ALCOHOL AND OTHER DRUGS OF ABUSE

It has been suggested that different neurotransmitter systems interact with specific relapse situations elicited by an alcohol priming dose, an alcohol-associated cue, or stress exposure in drug and alcohol dependence (Heinz 2002; Shalev et al. 2000). PET imaging has been most useful in such studies, because specific neurotransmitter systems can be identified. This model provides an example of how PET can be used to study the neurotransmitter mediation of important events related to addiction. One mainstay of the neurobiological research in alcohol and drug addiction is the observation that alcohol and all other drugs of abuse induce dopamine release in the ventral striatum, including the nucleus accumbens, and thus reinforce drug intake (Wise 1988).

Animal experiments also revealed that alcohol and drug-associated stimuli activate dopamine and endorphin release in the medial prefrontal cortex and the ventral striatum (Dayas et al. 2007; Di Chiara 2002; Shalev et al. 2000).

However, observing that a drug reinforces behavior does not necessarily imply that the drug effect is subjectively pleasant. Robinson and Berridge (1993) distinguished between hedonic or pleasant drug effects ("liking") and the craving for such a positive effect ("wanting"). They suggested that drug effects associated with drug liking are mediated by opioidergic neurotransmission in the ventral striatum including the nucleus accumbens, and that pleasurable effects during consumption of primary reinforcers such as food are also caused by endorphin release in this brain area (Berridge and Robinson 1998). Imaging studies are increasingly able to evaluate both of these limbs of the reward systems. Moreover, these hypotheses can be studied by using scores for drug craving as a proxy for wanting and scales measuring pleasure as a proxy for liking.

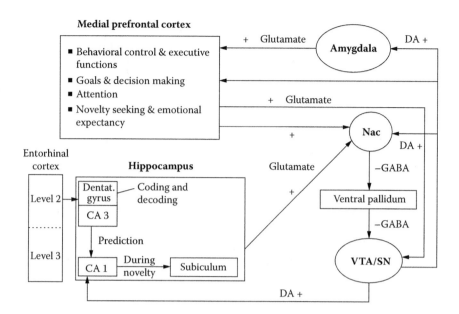

FIGURE 4.2 Model of a neuronal network that includes dopamine-related prediction of unexpected or novel reward and reward-associated stimuli: The discrepancy between expected and actual sensory information is calculated in the hippocampus (CA1) and activates dopaminergic neurons in the brainstem (VTA) via glutamatergic projections to the nucleus accumbens (ventral striatum). The VTA in turn modulates neuronal transmission in CA1 via an increased dopamine-release in the hippocampus and thus contributes to memory performance. The prefrontal cortex contributes to executive control functions and modulates the firing rate of dopaminergic neurons that project from the brain stem (VTA) to the nucleus accumbens (modified from Lisman and Grace 2005).

Dopamine release in the striatum is itself regulated by the hippocampus, which plays a major role in memory processes and reflects the appearance of novel stimuli (Lisman and Grace 2005). Stimulation of glutamatergic neurons in the hippocampus induces dopamine release in the ventral striatum of rats that previously consumed cocaine, and this glutamatergic-driven dopamine release can lead to a relapse in drug intake (Vorel et al. 2001). Therefore, experimental hippocampus stimulation reflects real-life situations in which drug-associated, contextual stimuli activate the hippocampus and trigger memories associated with previous drug use (Figure 4.2). The activation of this specific hippocampal circuit in turn activates dopamine neurons in the ventral tegmentum, which elicit dopamine release in the ventral striatum, thus facilitating new drug intake (Floresco et al. 2001). Indeed, both cocaine sensitization (Goto and Grace 2005b) and amphetamine sensitization (Lodge and Grace 2008) increase hippocampal drive of the nucleus accumbens, ultimately leading to increased responsivity of the dopamine (DA) system.

Berridge and Robinson (1998) suggested that a neurobiological correlate of "wanting" is phasic dopamine release in the ventral striatum, as shown by Schultz et al. (1997), which is not necessarily accompanied by positive feelings. Indeed, Schultz et al. had observed that the arrival of an unexpected reward elicits a burst of spikes in

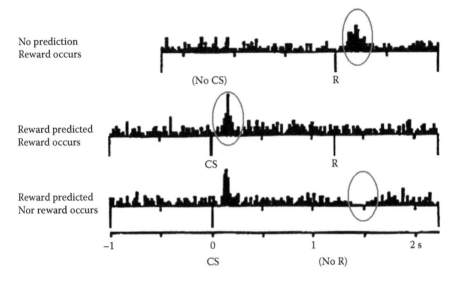

No prediction
Reward occurs

(No CS)   R

Reward predicted
Reward occurs

CS   R

Reward predicted
Nor reward occurs

−1   0   1   2 s

CS   (No R)

FIGURE 4.3   Reward-associated error signaling by short-term ("phasic") dopamine release (cf. Schulz et al. 1997). Top: An unexpected reward (banana pellets for rhesus monkeys), which was not predicted by previous stimuli, generates an error in reward prediction (unexpected reward) that is reflected in a short-term increase in dopamine firing. Middle: After learning that a previously neutral (now conditioned) stimulus (light) regularly predicts a reward, the surprising appearance of the conditioned stimulus reflects an error in reward prediction and generates a short-term increase in phasic dopamine firing rate. The reward itself is now completely predicted by the conditioned stimulus and does not elicit dopamine firing. Bottom: If a conditioned stimulus is not followed by the expected reward, an error in reward prediction occurs (unexpected lack of reward), which is reflected in a phasic decrease in dopamine firing.

dopaminergic neurons, which does not occur if this incident is predicted by a conditioned cue that temporarily predicts the subsequent arrival of a reward (Figure 4.3). The discharge of dopaminergic neurons occurs immediately after the presentation of such a conditioned cue and indicates the magnitude of the anticipated reward (Tobler et al. 2005). When the reward itself arrives as predicted (anticipated), it no longer elicits a dopamine discharge, because the received and the expected reward do not differ and there is thus no error of prediction, which is expressed by dopamine cell firing (Schultz et al. 1997) (Figure 4.3), as described below in a separate section. Robinson and Berridge (1998) suggested that phasic dopamine release not only represents this prediction error but also facilitates the allocation of attention toward salient, reward-indicating stimuli, which can thus motivate an individual to display a particular behavior in order to receive the reward. Dopamine thus contributes to the control of goal-directed behavior because it encodes the expected magnitude of a potential reinforcer and attributes incentive salient to reward-indicating stimuli. As a consequence, the nucleus accumbens acts as a "sensory-motor gateway," which controls the effects of salient environmental stimuli on prefrontal and limbic brain areas that regulate attention and motor output (Tobler et al. 2005).

PET can assess several aspects of dopaminergic function: D2 receptor number/ availability, dopamine synthesis rate, and dopamine release rate can be measured indirectly by evaluating the effect of amphetamine on D2 receptor occupation. Brain imaging studies among detoxified alcoholics that used PET and measured radio-ligand binding to central dopamine D2 receptors revealed a reduction in dopamine D2 receptor availability and sensitivity, which may reflect a homeostatic, compensatory downregulation of D2 receptors after chronic alcohol-associated dopamine release. Prospective studies showed that the degree of downregulation during early abstinence is associated with the subsequent relapse risk (Heinz et al. 1996; Volkow et al. 1996). Further PET studies showed that alcohol craving was specifically correlated with a reduced D2 receptor availability and a low dopamine synthesis capacity measured with F-DOPA PET in the ventral striatum including the nucleus accumbens (Heinz et al. 2004a, 2005c). During detoxification and early abstinence, dopamine dysfunction can further be increased, because alcohol-associated dopamine release is suddenly stopped and extracellular dopamine concentration decreases rapidly during detoxification (Rossetti et al. 1992). Stimulation of dopamine release with amphetamine during early abstinence confirmed that dopamine storage and stimulant-induced dopamine release is significantly reduced in detoxified alcohol dependents compared with healthy control subjects (Martinez et al. 2005). Altogether, these studies indicate that after alcohol detoxification, overall dopaminergic neurotransmission in the ventral striatum is reduced rather than sensitized. Therefore, it is quite unlikely that the presentation of alcohol-associated cues during early abstinence can cause a significant dopamine release that triggers reward craving or relapse.

Some animal studies confirmed that the presentation of drug- and alcohol-associated cues can lead to relapse even if no dopamine release can be observed in the ventral striatum (Shalev et al. 2002). However, these findings do not rule out that dopamine dysfunction plays a major role in the relapse risk of alcohol-dependent patients during early detoxification; rather, the assumed direction of dopamine dysfunction has to be reconsidered: According to the above-discussed animal experiments and human studies, a reduction in dopamine synthesis capacity, stimulant-induced dopamine release, and D2 receptor availability and sensitivity is associated with high alcohol craving, increased processing of alcohol-associated cues in the anterior cingulate and medial prefrontal cortex, and an increased relapse risk (Heinz et al. 2004a, 2005c). This hypothesis can be directly tested by simultaneously measuring dopaminergic neurotransmitters with PET and cue-reactivity with fMRI. Indeed, increased craving was associated with a low level of DA transmission. The decrease in DA levels could thereby serve as a drive for the individual to consume alcohol to boost low dopaminergic neurotransmission during early abstinence.

Imaging approaches can also be used to identify effects on the opioid component of the reward system. In human alcohol-dependent patients, an increase of mu-opiate receptors was observed in the ventral striatum, and high availability of mu-opiate receptors in the ventral striatum and medial prefrontal cortex correlated with the severity of alcohol craving (Heinz et al. 2005b). Naltrexone application in alcohol-dependent patients blocked alcohol-associated pleasurable feelings (the subjective "high"), indicating that drug liking may be associated with alcohol-induced activation of mu-opiate receptors and suggesting that a blockade of this interaction contributes to the relapse-preventing effects of naltrexone in alcoholism (O'Brien 2005; Volpicelli et al. 1995). Different human clinical studies showed that naltrexone treatment reduces the relapse risk of alcoholics (Srisurapanont and Jarusuraisin 2005; Streeton and Whelan 2001, but see Krystal et al. 2001).

4.3.1 NEUROTRANSMITTER DYSFUNCTION ASSOCIATED WITH ALTERED CUE REACTIVITY

To date only a handful of studies have examined the association between neurotransmitter dysfunction measured *in vivo* with PET or SPECT, cue-induced brain activation measured with fMRI, and the prospective relapse risk. In recently detoxified alcohol-dependent patients, the prospective relapse risk was associated with the extent of alcohol craving, which in turn was correlated with a reduced availability of dopamine D2 receptors in the ventral striatum of recently detoxified alcohol-dependent patients and with a low dopamine synthesis capacity measured with F-DOPA PET (Heinz et al. 2004a, 2005c). A low availability of dopamine D2 receptors in the ventral striatum of detoxified alcoholics was associated with increased functional activation of the anterior cingulate and adjacent medial prefrontal cortex elicited by alcohol-associated versus neutral pictures (Heinz et al. 2004a). The implicated brain areas had been associated with attribution of attention to salient stimuli (Fuster 1997), and cue-induced activation of these brain areas predicted the relapse risk in detoxified alcoholics (Grüsser et al. 2004).

4.4 BEHAVIORAL DESIGN OF STUDIES THAT MEASURE NEURONAL SYSTEMS ACTIVATED BY ALCOHOL-ASSOCIATED CUES

Functional MRI as well as PET or SPECT can be used to assess neuronal activation elicited by alcohol-associated cues via computation of the BOLD response associated with stimulus presentation. The key methodological issue in conducting such studies is not the imaging methodology that is used to conduct the study but the behavioral parameters used to engage neural circuits. In this next section, we review key studies that provide examples of the challenges involved in using drug stimuli themselves.

Brain-imaging studies have been used to describe neuronal networks responding to drugs of abuse, to link cue-induced brain activation with the prospective relapse risk (Braus et al. 2001; Drummond 2000; George et al. 2001; Grüsser et al. 2004), and to associate cue-induced brain activation with the effects of drugs that are supposed to block alcohol- or drug-specific effects.

These approaches that link brain activity to behavioral state have identified core regions that were activated in most studies (de Mendelssohn et al. 2004; Weiss 2005). These core regions include the following:

- the *orbitofrontal cortex (OFC)*, which is involved in the evaluation of reward and punishment (Myrick et al. 2004; Wrase et al. 2002)
- the *anterior cingulate* and adjacent *medial prefrontal cortex*, which is involved in attention and memory processes and in the encoding of the motivational value of stimuli (Grüsser et al. 2004; Heinz et al. 2004a; Myrick et al. 2004; Tapert et al. 2004)
- the *ventral striatum (including the nucleus accumbens)*, which connects motivational aspects of salient stimuli with motor reactions (Braus et al. 2001; Myrick et al. 2008; Wrase et al. 2002, 2007)
- the *dorsal striatum*, which has been implicated in habit formation and consolidates stimulus–reaction patterns (Grüsser et al. 2004; Modell and Mountz 1995)
- the *basolateral amygdala*, which specifies the emotional salience of stimuli and initiates conditioned and unconditioned approach and avoidance behavior (Schneider et al. 2001)

Similar brain areas are also activated when drug-specific stimuli were presented to patients addicted to other drugs of abuse, such as cocaine or opiates (de Mendelssohn et al. 2004; Weiss 2005).

However, these brain imaging studies revealed considerable interindividual variance in response to the presentation of alcohol-related stimuli, which may be due to differences in chosen populations, availability of alcohol, and motivation to remain sober as well as the design of the behavioral study.

4.4.1 Important Behavioral Parameters in Imaging Studies

Differences in experimental settings, stimulus selection, and presentation as well as in the chosen population can all influence brain activation by drug-related cues. The activation of brain areas appears to depend upon the sensory quality of the presented stimuli: For example, the fusiform gyrus is typically activated by visual, but not olfactory, cues (Braus et al. 2001). Another influencing factor is the state of detoxification and alcohol availability. The dorsolateral prefrontal cortex, for example, contributes to executive behavior control and was activated in acutely drinking patients when given a priming dose of alcohol (George et al. 2001).

The importance of the sensory stimuli used to activate neural circuits is illustrated in studies that use brain imaging to identify areas important in alcohol and drug craving. No consistent picture appears when the association between cue-induced activity in the above-described brain areas and subjective craving for alcohol is considered. For example, one study observed an association between the severity of craving and functional brain activation in the dorsal striatum (Modell and Mountz 1995); another one in the ventral striatum, orbitofrontal cortex, and anterior cingulate cortex (Myrick et al. 2004); and a third one in the subcallosal gyrus (Tapert et al. 2004). Other studies observed no significant correlation between alcohol craving and cue-induced brain activation, although the degree of cue-induced brain activation predicted relapse within the follow-up period of 3 months (Braus et al. 2001; Grüsser et al. 2004). The diverse nature of the applied stimuli may help to explain these different study results, especially since some studies used alcohol related pictures either with or without a sip of alcohol (Myrick et al. 2004; Grüsser et al. 2004), while other studies used alcohol-related words or odors (Schneider et al. 2001; Tapert et al. 2004). Words are expected to activate brain areas associated with language processing, but odors may more directly activate the amygdala and other limbic regions.

The duration of stimulus presentation may influence activation patterns elicited by the presentation of visual alcohol cues. When stimuli were presented not in a block design for up to 20 seconds but in a single event design, in which stimuli could only been seen for 750 ms, such briefly presented alcohol cues elicited increased brain activation in the prefrontal and cingulate cortex of detoxified alcohol-dependent patients (Heinz et al. 2007a). However, no significant correlation with the subsequent relapse risk was found.

The state of the subject is also critical. For example, in alcohol craving studies, the availability of alcohol may also play a role, since in some studies patients are not detoxified and thus able to consume alcohol, while in other studies patients participate in an inpatient treatment program, where relapse is not expected (Braus et al. 2001; Grüsser et al. 2004; Heinz et al. 2004a, 2007a; Wrase et al. 2002, 2007).

Difficulties in the assessment of subjective craving may also contribute to the inconsistencies in the literature regarding the association between alcohol urges and cue-induced brain activation. For example, craving can be assessed acutely or over a longer period, and it may vary depending on the availability of drugs; that is, it is usually higher when drugs are present (Wilson et al. 2004).

4.4.2   CUE-INDUCED BRAIN ACTIVATION AND THE PROSPECTIVE RELAPSE RISK

The sections above have described how to use imaging approaches to identify neural circuits involved in the processing of drug- and alcohol-associated cues as well as the elements of behavioral control that are critical to such approaches. Another important goal of most imaging studies is to provide insight into the relationships between the neural circuits that are activated and the behavioral responses. For example, identifying behavioral responses and their associated neuronal events that predict relapse is a major goal of much such research.

Three major events have been identified that can induce a relapse in detoxified patients suffering from alcohol and other drug addictions that can provide methodological tools to experimentally assess cue-induced neuronal responses: exposure to stress, exposure to a priming dose of alcohol, and exposure to stimuli that have regularly been associated with alcohol intake (cues). These events have been implicated in studies trying to identify factors eliciting relapse (Adinoff 2004; Berridge and Robinson 1998; Breese et al. 2005; Cooney 1997; Di Chiara and Bassareo 2007; Everitt and Robbins 2005). Although a series of brain imaging studies investigated the association between brain activation elicited by alcohol-associated stimuli and alcohol craving, only very few studies assessed the clinical relevance of these activation patterns, that is, to what extent cue-induced brain activation can predict the prospective relapse risk after detoxification. One pilot study with alcohol-dependent patients revealed that alcohol cues elicited increased activation of the ventral striatum and visual association centers in detoxified alcoholics. It showed that patients who suffered from multiple relapses during their previous course of disease and who relapsed rather quickly after detoxification showed a stronger cue-induced activation of the ventral striatum than patients who had previously managed to abstain from alcohol for longer periods of time and who also abstained during the six-month follow-up period (Braus et al. 2001). This finding was confirmed by a study of Grüsser et al. (2004), which also observed in a rather small sample that subsequently relapsing versus abstaining patients displayed an increased brain activation elicited by visual alcohol-associated stimuli in the anterior cingulate and adjacent medial prefrontal cortex and in the central striatum. The central striatum, which was activated in the study of Grüsser et al., is a part of the dorsal striatum; it has been suggested that the dorsal striatum is crucial for habit learning, for example, for the learning of automated responses, and may thus contribute to the compulsive character of dependent behavior. Studies can therefore distinguish between neuronal correlates of habitual drug intake (i.e., without conscious craving) and drug intake following drug urges.

The importance of habitual drug intake is reflected in the assessment of brain areas associated with cue-induced activation: Cue-induced conscious craving may preferentially elicit dopamine release in more dorsal striatal structures among subjects suffering from drug addiction (Volkow et al. 2006; Wong et al. 2006). This transition from a predominate response in the ventral to the dorsal striatum can reflect a transition from a reward-driven phenomenon associated with activation of the ventral striatum to a stimulus-response habit formation depending on dorsal striatal activation (Berke and Hyman 2000). In such automated responses, the experience of actual reward may play a lesser role. In accordance with this hypothesis, Robbins and Everitt (2002) proposed that the initial reinforcing effects of drugs of abuse may depend on activation of the ventral striatum, while transitions from initial drug taking into habitual drug seeking may be associated with activation of more dorsal striatal regions. Studies with PET also suggested that among addicted individuals, drug cues tend to preferentially release dopamine in the dorsal striatum and putamen (Volkow et al. 2006; Wong et al. 2006). Activation of the dorsal striatum may thus be associated with habit formation, and indeed, many patients describe their relapse in terms of automated actions and do not remember to have experienced craving before relapse (Tiffany 1990).

The use of drug versus nondrug stimuli can markedly influence the brain areas that are activated. Another study observed that very briefly presented alcohol cues elicited increased brain activation in the prefrontal and cingulate cortex of detoxified alcohol-dependent patients (Heinz et al. 2007a); however, no significant correlation with the subsequent relapse risk was found. This was different when brain activation elicited by affectively positive stimuli was assessed: alcohol-dependent patients who displayed increased activation of the ventral striatum following presentation of pleasant stimuli showed a subsequently *reduced* relapse risk. A study of the group of Markus Heilig suggested that drugs that increase neuronal activation elicited by affectively positive stimuli may reduce the relapse risk of alcohol-dependent patients

(George et al. 2008). Indeed, one possibility is that the affectively positive stimuli may be reinstating the ventral striatal association between stimulus and reward that had been supplanted by dorsal striatal habit formation.

Relapse may thus occur when alcohol-associated, but not affectively positive, stimuli activate the ventral and dorsal striatum and anatomically closely associated brain areas such as the anterior cingulate cortex. Relapse risk may further be increased if brain areas associated with executive behavior control such as the dorsal lateral prefrontal cortex are impaired in chronic alcohol and drug dependence. One study in methamphetamine-dependent patients confirmed this hypothesis and suggested that the subsequent relapse risk can be predicted by activation patterns elicited during a decision-making task in the posterior cingulate and temporal cortex and in the insula (Paulus et al. 2005). However, so far no study has assessed the association between dysfunction of decision making and the prospective relapse risk in alcohol dependence.

4.5   ALCOHOL CRAVING: A LEARNED RESPONSE?
HOW TO ASSESS THE IMPACT OF LEARNING

Drug addictions such as alcohol dependence are characterized by criteria that include the development of tolerance to drug effects, withdrawal symptoms upon cessation of drug intake, craving for the drug of abuse, and reduced control of drug intake (American Psychiatric Association 1994; World Health Organization 1992). Koob (2003) suggested that tolerance can be understood as an adaptation of the brain to chronic drug intake, which results in a new homeostatic balance. This balance is thought to be disturbed when alcohol intake is suddenly interrupted, and the resulting homeostatic imbalance can clinically manifest as withdrawal symptoms. To methodologically assess these questions, neuroreceptor alterations (adaptations) can be studied during drug intake and compared with the state of acute detoxification and prolonged abstinence. Indeed, it has been observed that the sedative effects of alcohol and other drugs of abuse are mediated by stimulation of GABAergic neurotransmission; alcohol has also been shown to inhibit glutamatergic neurotransmission, thus effectively interfering with excitatory brain activation (Krystal et al. 2006; Tsai et al. 1995). Once alcohol intake is suddenly stopped during withdrawal, GABAergic receptors are no longer activated by alcohol, and upregulated glutamate receptors are no longer functionally inhibited by direct ethanol effects. The imbalance may result in withdrawal symptoms, even as severe as epileptic seizures (Krystal et al. 2006; Tsai et al. 1995). It is obvious that patients can relapse while experiencing such aversive and potentially life-threatening withdrawal symptoms. However, it has also been observed that such withdrawal symptoms can manifest even when acute detoxification and withdrawal symptoms have ceased for a considerable amount of time. These can be assessed with questionnaires such as the Clinical Institute Withdrawal Assessment for Alcohol Scale (CIWA-Ar) that measure the severity of withdrawal symptoms preceding relapse in detoxified patients (Heinz et al. 2003).

It has been suggested that manifestation of withdrawal symptoms during abstinence can be explained in part as a conditioned reaction elicited by conditioned stimuli, that is, cues that have been regularly associated with alcohol intake, which can elicit conditioned responses such as craving for the rewarding effects of alcohol or neurobiological effects such as a counteradaptive neuronal response that is aimed at counteracting the expected drug effect (Siegel 1975; Siegel et al. 1982; Wikler 1948). Indeed, it has been suggested that in alcohol dependence, contextual cues that characterize situations in which alcohol is available can act as conditioned stimuli that trigger counteradaptive alterations in neurotransmitter systems such as the glutamate and GABA systems, inducing increased glutamatergic and decreased GABAergic neurotransmission. If alcohol intake does not occur as expected, the resulting hyperexcitation can manifest as withdrawal symptoms and trigger relapse (Verheul et al. 1999). It has been suggested that in such situations, patients may experience craving for alcohol's sedative effects, which are motivated by the desire to relieve the aversive experience of conditioned withdrawal. In a clinical study, about 30% of all alcohol-dependent patients described that their last relapse was preceded by a sudden manifestation of withdrawal symptoms, which often occurred long after acute detoxification and which were triggered by being exposed to previously typical drinking situations (Heinz et al. 2003).

Conditioned stimuli may not only trigger neurobiological effects that are meant to counteract the expected drug response, as described above for opiates; but it has also been suggested that drugs of abuse, particularly those with a strong dopaminergic component, may also trigger expectation of the motivation or reinforcing effects of the drug of abuse (Heinz et al. 2003; Stewart et al. 1984; Verheul et al. 1999). Thus, when a rat is given a stimulant within a particular context (e.g., their home cage), the rat will show sensitization, or increased response to the same dose of the drug, when it is tested in the same context in which it had been administered (Vezina et al. 1989; Badiani et al. 2000; Crombag et al. 2000). A learning theory suggests that originally neutral stimuli can be associated with alcohol's affectively positive effects (unconditioned response, UCR) or with a compensatory, homeostatic counteradaptive process (Koob 2003; Siegel 1999). These cues thus become conditioned stimuli (CS), which have been associated with the subjectively pleasant or counteradaptive drug effects. Conditioned stimuli can thus elicit an urge or "craving" as a conditioned response (CR) for the affectively positive effects of alcohol or other drugs of abuse or for the alleviation of an aversive, homeostatic counteradaptive process (Bigelow 2001; Drummond 2000; Verheul et al. 1999) (Figure 4.4). It is important to note that such formerly neutral stimuli, which are associated with the effects of a drug of abuse, can be both internal or external cues, for example, the context or environment that characterized former alcohol consumption or internal stimuli such as feelings of loneliness or memories of conflict situations, which had previously been associated with excessive alcohol consumption (Drummond 2000; Heinz et al. 2003; Verheul et al. 1999). Small quantities of the substance itself can also trigger craving via conditioned responses. Therefore some groups studied—for example, nondetoxified alcohol-dependent patients—were given a sip of alcohol to elicit craving (Myrik

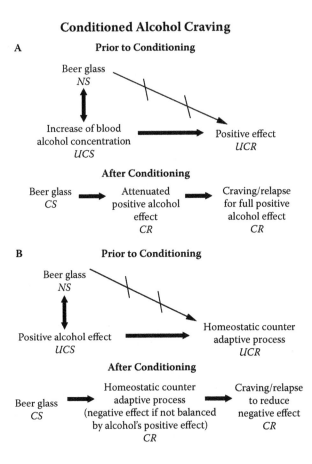

FIGURE 4.4 Model of conditioned alcohol craving: A previously neutral stimulus (e.g., beer glass) can become associated with (a) an increase in blood alcohol concentration that elicits an affectively positive response (UCR). The CS can then elicit an attenuated positive effect and craving for the "full" affectively positive effect of alcohol. Furthermore, the positive effect of alcohol itself (b) can act as an unconditioned stimulus (UCS) that elicits a homeostatic counteradaptive process as an unconditioned response (UCR). If the CS is presented without subsequent alcohol intake, the counteradaptive process is experienced as aversive, and craving for alcohol emerges to reduce this aversive state.

et al. 2004). These studies show that it is methodogically important to distinguish between different stimuli that elicit craving (alcohol sips, pictures, or other cues) and the associated urges (i.e., craving for positive drug effects or for the alleviation of aversive withdrawal symptoms).

4.5.1 CLINICAL STUDIES TRYING TO LINK CRAVING AND RELAPSE: IDENTIFYING CRAVING

In alcohol dependence, the empirical connection between craving for alcohol and the following relapse risk is not well explored. Although animal experiments strongly support the hypothesis that conditioned drug reactions are involved in the development and maintenance of addictive behavior and relapse (Di Chiara 2002; Robbins and Everitt 2002; Robinson and Berridge 1993), human studies often did not find a positive correlation between subjective, or conscious, alcohol craving and relapse (Drummond and Glautier 1994; Grüsser et al. 2004; Junghanns et al. 2005; Kiefer et al. 2005; Litt et al. 2000; Rohsenow et al. 1994). However, some other studies did observe such relationship (Bottlender and Soyka 2004; Cooney et al. 1997; Heinz et al. 2005c; Ludwig and Wikler 1974; Monti et al. 1990). In contrast to these rather heterogeneous results, changes in physiological parameters elicited by alcohol-associated cues, including neuronal activation measured with functional imaging, seem to be more closely connected to relapse (Abrams et al. 1988; Braus et al. 2001; Drummond and Glautier 1994; Grüsser et al. 2004; Rohsenow et al. 1994). Assessing such physiological responses may therefore be more promising to achieve reliable results.

One explanation for the poor correlation between alcohol craving and the prospective relapse risk may be due to the fact that reactions such as conscious craving to alcohol and alcohol-associated cues differ with respect to different levels of description (subjective, motor, and physiological responses), even if these reactions manifest at the same time point after cue exposure (Figure 4.5). Tiffany (1990) suggested that conscious craving only occurs if the automatic process of drug intake is interrupted. Whenever this is not the case, conditioned stimuli may trigger dramatic drug intake even in the absence of conscious urges for drugs of abuse. This theoretical framework suggests that behavioral patterns associated with drug intake can be activated in the absence of conscious correlates such as craving. This hypothesis is supported by studies that identified subcortical brain areas associated with motor behavior such as the striatum and that linked this cue-induced brain activation and relapse (Grüsser et al. 2004). It is unlikely that striatal activation per se is associated with conscious cognitive correlates; thus automated drug intake may occur against a conscious decision to remain abstinent.

Furthermore, heterogeneous research methods, settings and samples, and confounding effects of nicotine abuse, particularly in alcohol dependence, may explain inhomogeneous data concerning the association between craving and relapse as

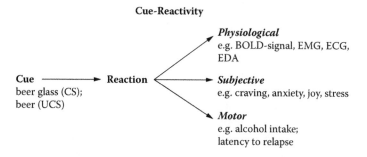

FIGURE 4.5 Measuring of "cue-reactivity referring" to Drummond (2000). The reactions triggered by drug- or alcohol-associated stimuli ("cues") can be measured on different levels, on which they are distinguished differently depending on the individual disposition and learning history.

mentioned before. For example, cue-induced craving and enhanced drug-like arousal were not associated in abstaining alcoholics, indicating that the psychological and physiological levels are dissociated (Breese et al. 2005). Likewise, Weiss et al. (2003) showed in a study with cocaine-addicted patients that an effective psychosocial treatment aimed at the reduction of drug craving helped patients to stay abstinent in spite of the persistence of subjectively high craving. It has even been suggested that a conscious sensation of alcohol craving can serve as a warning sign, which helps patients to look for help and thus to maintain abstinence (Drummond and Glautier 1994; Monti et al. 1990). However, craving appears to induce relapse if it occurs in stressful situations (Breese et al. 2005; Cooney et al. 1997). Assessment of the specific context that elicits drug urges and relapse thus appears very important.

4.5.2 Computational Models of Phasic Dopamine Release

PET measures of dopamine synthesis capacity and D2 receptor availability examine static, traitlike, rather than state, markers and thus do not measure phasic dopamine release. Such combined imaging studies, for example, F-DOPA and raclopride, revealed a combination of both a low presynaptic dopamine synthesis capacity and a low availability of dopamine D2 receptors, which suggest that during early detoxification and abstinence, dopamine dysfunction is characterized by a reduced rather than a sensitized dopamine production and receptor state. How can such a reduced function of dopaminergic neurotransmission in the ventral striatum be associated with increased processing of alcohol-associated salient stimuli?

One possible explanation is given when the work of Schultz et al. (1997) is taken into account. As mentioned above, Schultz and co-workers showed that phasic alterations in dopamine release are elicited when reward or a conditioned stimulus associated with reward is unpredictedly encountered. It has been suggested that such dopamine activations are required to learn the motivational salience of new stimulus reward associations. However, Schultz and co-workers also showed that dopaminergic neurotransmission may be involved in the process of unlearning established associations. Precisely, it was observed that a transient cessation of dopamine neuron firing (phasic "dip") of dopamine self-firing occurs whenever a conditioned stimulus is not followed by the anticipated reward (Figure 4.3). In this case, the received reward is lower than the anticipated reward, and the difference between these two values is negative, reflected in a reduction of dopamine cell firing. This adaptation of phasic dopamine cell firing thus acts as an error-detection signal, which indicates unexpected arrival of salient new stimuli and of surprising rewards, and that also reflects the nonappearance of an anticipated reinforcer.

Dopamine dysfunction during early abstinence can be detected with PET approaches. For example, low dopamine synthesis, reduced stimulant-induced dopamine release, and D2 receptor availability have been identified in the ventral striatum of detoxified alcoholics (Heinz et al. 2004a, 2005c; Martinez et al. 2005) and can interfere with this dopamine-dependent signaling of an error in reward expectation. Individually well-known cues such as alcohol-associated stimuli, which are no longer followed by alcohol reward once patients have stopped alcohol intake, can thus fail to elicit an adequate neuronal computational process when the expected reward does not occur (i.e., the phasic dip in dopamine cell firing). The alcohol-dependent individual may thus have problems unlearning previously well established associations between an alcohol cue and a conditioned response such as drug craving or drug seeking. Instead, it may be rather difficult for alcohol-dependent patients to divert attention away from conditioned cues that have been well learned to signal the availability of alcohol, potentially via glutamate-dependent long-term potentiation of a circuitry that includes the ventral hippocampus and ventral striatum and contributes to perseverative behavior (Goto and Grace 2005b). Dopamine dysfunction may thus directly interfere with a phasic, dopamine-dependent error signal, which indicates that alcohol-associated cues are no longer followed by reward.

Dopamine release modulated by drugs and cue-induced craving can potentially modulate ventral striatal inputs in a manner consistent with drug-seeking behavior. The prefrontal cortical and hippocampal inputs to the ventral striatum are differentially regulated by dopamine, with D1 receptors potentiating hippocampal inputs and D2 receptors attenuating prefrontal cortical inputs (Goto and Grace 2005a). The ventral hippocampus has been described as an area important in context-dependent behaviors, and thus in keeping focused on a task (Jarrard 1995; Sharp 1999; Grace et al. 2007). In contrast, one function of the prefrontal cortex is in driving behavioral flexibility (Ragozzino 2007). Administration of a drug or presentation of a reinforcer would activate dopamine neuron firing and increase dopamine transmission; this would result in facilitation of hippocampal input (i.e., staying on task) while inhibiting prefrontal cortical input (i.e., prevent shifting in strategies). In contrast, if the cue is not followed by a reinforcer, the subsequent decrease in DA neuron firing would produce the opposite effect: an attenuation of hippocampal inputs and facilitation of prefrontal inputs, which would favor shifting of behavioral strategies (Goto and Grace 2008). Thus, stimuli that result in unexpected rewards would favor continuation of that response, while events that fail to produce expected rewards would favor a shifting of strategies. This hypothesis could directly be studied by combining PET and fMRI studies to quantify, for example, dopamine displacement of a radioligand and cue-induced brain activation.

PET imaging can also be used to identify the correlations between neurotransmitter dysfunction induced by chronic drug use and normal behavioral responses like those to natural reinforcers. If dopamine dysfunction interferes with phasic changes in dopaminergic neurotransmission among detoxified alcoholics, such patients should also have problems in attributing salience to newly learned conditioned stimuli, particularly when they are presented unexpectedly and indicate the availability of reward. As a support of this assumption, a reduced functional activation of the ventral striatum was observed among alcohol-dependent patients who were recently detoxified and confronted with newly learned, previously abstract stimuli that indicate the availability of reward (Wrase et al. 2007). The reduced activation of the ventral striatum during the presentation of such conditioned, reward-indicating stimuli correlated with the severity of alcohol craving and was not explained by differences in task performance or mood between alcoholics and control subjects.

In early detoxification and abstinence, the patient's motivation to experience new and potentially rewarding situations may thus be impaired by dysfunction of the ventral striatum, which fails to be adequately activated by new reward-indicating stimuli (Wrase et al. 2007). In the study of Wrase et al. (2007), the same alcohol-dependent patients displayed an increased activation of the ventral striatum upon confrontation with alcohol-associated cues, and this activation of the ventral striatum was positively correlated with the severity of alcohol urges. A reduced activation of the brain reward system toward new, reward-indicating stimuli and an increased response to drug cues are observations that support the hypotheses that alcohol and other drugs of abuse "hijack" a dysfunctional reward system, which fails to adequately process conventional, primary reinforcers such as food or sex, but which tends to respond too strongly to drug-associated stimuli (Volkow et al. 2004). This finding may also help to explain why it can be difficult to motivate detoxified alcoholics to replace alcohol by other reinforcers such as new hobbies or social interactions; since the neuronal responses to alcohol-associated cues are increased while an activation elicited by new, reward-indicating stimuli is reduced, patients appear to experience difficulties when trying to divert attention away from alcohol-associated cues that signal the availability of alcohol and its dopamine-stimulating pharmacological effect (Di Chiara 2002; Di Chiara and Bassareo 2007).

4.6   USING IMAGING TO MAKE CLINICAL DIAGNOSES

So far imaging cannot be applied to facilitate diagnosis of alcohol dependence. However, functional imaging studies can help to identify patients who display strong physiological cue reactivity and who are thus at risk to suffer a relapse when confronted with alcohol-associated cues. Currently, imaging techniques such as fMRI are too expensive for broad clinical use. However, less complicated techniques, such as the affect-modulated startle response, which assesses physiological responses to affectively positive or negative as well as drug-associated stimuli, may help to identify patients with a particularly high relapse risk. Physiological markers such as the startle response that reflect an appetitive response toward drug cues are particularly important because many patients deny alcohol craving during the presentation of alcohol cues but show strong appetitive reactions when assessed with the startle response (Heinz et al. 2003).

Patients who suffer from strong cue-reactivity and a high risk of relapse may specifically profit from certain psychotherapeutic treatments such as cue exposure. Cue exposure has repeatedly been investigated in therapeutic studies; however, to date this treatment approach does not seem to yield significantly better results than standard therapy with cognitive behavioral and supporting interventions (Kavanagh et al. 2004; Löber et al. 2006). However, cue exposure may work best among patients with strong neuronal responses to alcohol cues, and identification of such patients may thus help to treat this subgroup of patients with greater success.

Brain imaging may also be used to assess the effects of additive pharmacotherapy on cue-induced neuronal activation patterns. Hermann and co-workers (2006) showed in one pilot study that cue-induced activation of the thalamus is blocked by acute application of amisulpride in detoxified alcoholics. The group of Anton and co-workers (Myrick et al. 2008) was able to show that cue-induced activation of limbic brain areas is reduced by naltrexone and by the combination of naltrexone and ondansetron in detoxified alcoholics, and alterations in the response to affective cues have been suggested to predict relapse (Heinz et al. 2007a) and may help to identify new pharmacological treatment strategies such as modulation of central stress responses (George et al. 2008).

4.7 SUMMARY AND OUTLOOK

Neurobiological research on different mechanisms of relapse may help to identify individually vulnerable patterns and can be used to adapt treatment toward the individual needs of patients. Moreover, neuroscientific research may help to reduce the stigma of addiction, because it shows that alcohol-dependent patients do not suffer from "weak willpower" or "bad intentions," as suggested early during the 20th century. Instead, brain imaging studies suggest that alcohol-associated cues activate limbic brain areas, which appear to be in part genetically influenced and to respond in an automated, habitlike manner that is hardly influenced by conscious intentions (Heinz et al. 2005a, 2003). Indeed, cue-induced brain activation predicted the relapse risk of alcohol-dependent patients better than conscious craving, a finding that is not surprising when taking into consideration that activation of brain areas such as the striatum is hardly associated with conscious experiences. Therefore, it seems plausible that patients experience drug-craving and drug-seeking behavior "against their own conscious will"; these should not be blamed for their behavior, but instead be treated with the same respect as other patients in the health care system.

ACKNOWLEDGMENT

We thank Professor Fred Rist (Institute of Psychology I: Psychological Diagnostics and Clinical Psychology, University of Münster, Germany) for his expert counseling in describing learning theories of addiction. This work was supported by the Deutsche Forschungsgemeinschatf (HE 2597/4-3m HE 2597/7-3) and the BMBF (NGFN Plus 01GS08159).

REFERENCES

Abrams DB, Monti PM, Carey KB, Pinto RP, Jacobus SI (1988) Reactivity to smoking cues and relapse—two studies of discriminant validity. *Behav Res Ther* 26:225–233.

Adinoff B (2004) Neurobiologic processes in drug reward and addiction. *Harv Rev Psychiatry* 12:305–320.

American Psychiatric Association (1994) *Diagnostic and Statistical Manual of Mental Disorders, 4th ed.* Washington, D.C.: American Psychiatric Press.

Badiani A, Oates MM, Fraioli S, Browman KE, Ostrander MM, Xue CJ, Wolf ME, Robinson TE (2000) Environmental modulation of the response to amphetamine: dissociation between changes in behavior and changes in dopamine and glutamate overflow in the rat striatal complex. *Psychopharmacology* (Berl) 151:166–174.

Berke JD, Hyman SE (2000) Addiction, dopamine, and the molecular mechanisms of memory. *Neuron* 25: 515–532.

Berridge KC, Robinson TE (1998) What is the role of dopamine in reward: hedonic impact, reward learning, or incentive salience? *Brain Res Brain Res Rev* 28:309–369.

Bigelow GE (2001) An operant behavioral perspective on alcohol abuse and dependence. In: *International handbook of alcohol dependence and problems*. Heather N, Peters TJ, Stockwell T (Eds.). 299–315. John Wiley & Sons.

Boothby LA, Doering PL (2005) A camprosate for the treatment of alcohol dependence. *Clin Ther* 27:695–714.

Bottlender M, Soyka M (2004) Impact of craving on alcohol relapse during, and 12 months following, outpatient treatment. *Alcohol Alcohol* 39:357–361.

Braus DF, Wrase J, Grüsser S, Hermann D, Ruf M, Flor H, Mann K, Heinz A (2001) Alcohol-associated stimuli activate the ventral striatum in abstinent alcoholics. *J Neural Transm* 108:887–894.

Breese GR, Chu K, Dayas CV, Funk D, Knapp DJ, Koob GF, Lê DA, O'Dell LE, Overstreet DH, Roberts AJ, Sinha R, Valdez GR, Weiss F (2005) Stress enhancement of craving during sobriety: a risk for relapse. *Alcohol Clin Exp Res* 29:185–195.

Carter B, Tiffany S (1999) Meta-analysis of cue-reactivity in addiction research. *Addiction* 94:327–340.

Cooney NL, Litt MD, Morse PA, Bauer LO, Gaupp L (1997) Alcohol cue reactivity, negative-mood reactivity, and relapse in treated alcoholic men. *J Abnorm Psychol* 106:243–250.

Crombag HS, Badiani A, Maren S, Robinson TE (2000) The role of contextual versus discrete drug-associated cues in promoting the induction of psychomotor sensitization to intravenous amphetamine. *Behav Brain Res* 116:1–22.

Dayas CV, Liu X, Simms JA, Weiss F (2007) Distinct patterns of neural activation associated with ethanol seeking: effects of naltrexone. *Biol Psychiatry* 61(8):979–989.

De Mendelssohn A, Kasper S, Tauscher J (2004) Neuroimaging in substance abuse disorders. *Nervenarzt* 75:651–662.

Di Chiara G (2002) Nucleus accumbens shell and core dopamine: differential role in behavior and addiction. *Behav Brain Res* 137(1–2):75–114.

Di Chiara G, Bassareo V (2007) Reward system and addiction: what dopamine does and doesn't do. *Curr Opin Pharmacol* 7(1):69–76.

Drummond DC (2000) What does cue-reactivity have to offer clinical research? *Addiction* 95: S129–S144.

Drummond DC, Glautier S (1994) A controlled trial of cue exposure treatment in alcohol dependence. *J Consult Clin Psychol* 62:809–817.

Everitt BJ, Robbins TW (2005) Neural systems of reinforcement for drug addiction: from actions to habits to compulsion. *Nat Neurosci* 8:1481–1489.

Fisher RE, Morris ED, Alpert MN, Fischman AJ (1995) In vivo imaging of neuromodulatory synaptic transmission using PET: a review of relevant neurophysiology. *Hum Brain Map* 3:24–34.

Floresco SB, Todd CL and Grace AA (2001) Glutamatergic afferents from the hippocampus to the nucleus accumbens regulate activity of ventral tegmental area dopamine neurons. *J Neurosci* 21:4915–4922.

Fuster JM (1997) *The prefrontal cortex: anatomy, physiology, and neuropsychology of the frontal lobe*. Philadelphia: Lippincott-Raven.

George DT, Gilman J, Hersh J, Thorsell A, Herion D, Geyer C, Peng X, Kielbasa W, Rawlings R, Brandt JE, Gehlert DR, Tauscher JT, Hunt SP, Hommer D, Heilig M (2008) Neurokinin 1 receptor antagonism as a possible therapy for alcoholism. *Science* 319(5869):1536–1539.

George MS, Anton RF, Bloomer C, Teneback C, Drobes DJ, Lorberbaum JP, Nahas Z, Vincent DJ (2001) Activation of prefrontal cortex and anterior thalamus in alcoholic subjects on exposure to alcohol-specific cues. *Arch Gen Psychiatry* 58:345–352.

Goto Y, Grace AA (2005a) Dopaminergic modulation of limbic and cortical drive of nucleus accumbens in goal-directed behavior. *Nature-Neuroscience* 8:805–812.

Goto Y, Grace AA (2005b) Dopamine-dependent interactions between limbic and prefrontal cortical synaptic plasticity in the nucleus accumbens: Disruption by cocaine sensitization. *Neuron* 47:255–266.

Goto Y, Grace AA (2008) Limbic and cortical information processing in the nucleus accumbens. *Trends in Neurosciences* 31:552–558.

Grace AA, Floresco SB, Goto Y, Lodge DJ (2007) The regulation of dopamine neuron firing and the control of goal-directed behaviors. *Trends in Neurosciences* 30:220–227.

Grüsser SM, Wrase J, Klein S, Hermann D, Smolka MN, Ruf M, Weber-Fahr W, Flor H, Mann K, Braus DF, Heinz A (2004) Cue-induced activation of the striatum and medial prefrontal cortex is associated with subsequent relapse in abstinent alcoholics. *Psychopharmacology* (Berl) 175:296–302.

Gusnard DA, Raichle ME (2001) Searching for a baseline: Functional imaging and the resting human brain. *Nat Rev Neurosci* 2:685–694.

Heinz A (2002) Dopaminergic dysfunction in alcoholism and schizophrenia—psychopathological and behavioral correlates. *Eur Psychiat* 17:9–16.

Heinz A, Braus DF, Smolka MN, Wrase J, Puls I, Hermann D, Klein S, Grüsser SM, Flor H, Schumann G, Mann K, Buchel C (2005a) Amygdala-prefrontal coupling depends on a genetic variation of the serotonin transporter. *Nat Neurosci* 8:20–21.

Heinz A, Dufeu P, Kuhn S, Dettling M, Gräf K, Kürten I, Rommelspacher H, Schmidt LG (1996) Psychopathological and behavioral correlates of dopaminergic sensitivity in alcohol dependent patients. *Arch Gen Psychiatry* 53:1123–1128.

Heinz A, Jones DW, Zajicek K, Gorey JG; Juckel G, Higley JD, Weinberger DR (2004b) Depletion and restoration of endogenous monoamines affects β-CIT binding to serotonin but not dopamine transporters in non-human primates. *J Neural Transm* 68 (Suppl):29–38.

Heinz A, Löber S, Georgi A, Wrase J, Hermann D, Rey ER, Wellek S, Mann K (2003) Reward craving and withdrawal relief craving: assessment of different motivational pathways to alcohol intake. *Alcohol Alcohol* 38:35–39.

Heinz A, Reimold M, Wrase J, Hermann D, Croissant B, Mundle G, Dohmen BM, Braus DF, Schumann G, Machulla HJ, Bares R, Mann K (2005b) Correlation of stable elevations in striatal mu-opioid receptor availability in detoxified alcoholic patients with alcohol craving–a positron emission tomography study using carbon 11-labeled carfentanil. *Arch Gen Psychiatry* 62:57 64.

Heinz A, Saunders RC, Kolachana BS, Jones DW, Gorey JG, Bachevalier J, Weinberger DR (1999) Striatal dopamine receptors and transporters in monkeys with neonatal temporal limbic damage. *Synapse* 32(2):71–9.

Heinz A, Siessmeier T, Wrase J, Buchholz HG, Gründer G, Kumakura Y, Cumming P, Schreckenberger M, Smolka MN, Rösch F, Mann K, Bartenstein P (2005c) Correlation of alcohol craving with striatal dopamine synthesis capacity and D-2/3 receptor availability: a combined [18F]DOPA and [18F]DMFP PET study in detoxified alcoholic patients. *Am J Psychiatry* 162:1515–1520.

Heinz A, Siessmeier T, Wrase J, Hermann D, Klein S, Grüsser SM, Flor H, Braus DF, Buchholz HG, Gründer G, Schreckenberger M, Smolka MN, Rösch F, Mann K, Bartenstein P (2004a) Correlation between dopamine D-2 receptors in the ventral striatum and central processing of alcohol cues and craving. *Am J Psychiatry* 161:1783–1789.

Heinz A, Smolka MN, Braus DF, Wrase J, Beck A, Flor H, Mann K, Schumann G, Büchel C, Hariri AR, Weinberger DR (2007b) Serotonin transporter genotype (5-HTTLPR): effects of neutral and undefined conditions on amygdala activation. *Biol Psychiatry* 61(8):1011–1014.

Heinz A, Wrase J, Kahnt T, Beck A, Bromand Z, Grusser SM, Kienast T, Smolka MN, Flor H, Mann K (2007a) Brain activation elicited by affectively positive stimuli is associated with a lower risk of relapse in detoxified alcoholic subjects. *Alcohol Clin Exp Res* 31(7):1138–1147.

Hermann D, Smolka MN, Wrase J, Klein S, Nikitopoulos J, Georgi A, Braus DF, Flor H, Mann K, Heinz A (2006) Blockade of cue-induced brain activation of abstinent alcoholics by a single administration of amisulpride as measured with fMRI. *Alcohol Clin Exp Res* 30(8):1349–1354.

Jarrard LE (1995) What does the hippocampus really do? *Behav Brain Res* 71:1–10.

Junghanns K, Tietz U, Dibbelt L, Kuether M, Jurth R, Ehrenthal D, Blank S, Backhaus J (2005) Attenuated salivary cortisol secretion under cue exposure is associated with early relapse. *Alcohol Alcohol* 40:80–85.

Kavanagh DJ, Andrade J, May J (2004) Beating the urge: implications of research into substance-related desires. *Addict Behav* 29:1359–1372.

Kiefer F, Helwig H, Tarnaske T, Otte C, Jahn H, Wiedemann K (2005) Pharmacological relapse prevention of alcoholism: clinical predictors of outcome. *Eur Addict Res* 11:83–91.

Koob GF (2003) Alcoholism: allostasis and beyond. *Alcohol Clin Exp Res* 27(2):232–243.

Krystal JH, Cramer JA, Krol WF, Kirk GF, Rosenheck RA, Veterans Affairs Naltrexone Cooperative Study 425 Group (2001) Naltrexone in the treatment of alcohol dependence. *N Engl J Med* 345(24):1734–1739.

Krystal JH, Staley J, Mason G, Petrakis IL, Kaufman J, Harris RA, Gelernter J, Lappalainen J (2006) Gamma-aminobutyric acid type A receptors and alcoholism: intoxication, dependence, vulnerability, and treatment. *Arch Gen Psychiatry* 63(9):957–968.

Kumakura Y, Cumming P, Vernaleken I, Buchholz HG, Siessmeier T, Heinz A, Kienast T, Bartenstein P, Gründer G (2007) Elevated [18F]fluorodopamine turnover in brain of patients with schizophrenia: an [18F]fluorodopa/positron emissiontomography study. *J Neurosci* 27:8080–8087.

Kwong KK, Belliveau JW, Chesler DA, Goldberg IE, Weisskoff RM, Poncelet BP, Kennedy DN, Hoppel BE, Cohen MS, Turner R, Cheng H, Brady TJ, Rosen BR (1992) Dynamic magnetic resonance imaging of human brain activity during primary sensory stimulation. *PNAS* 89:5675–5679.

Laruelle M (2000) Imaging synaptic neurotransmission with in vivo binding competition techniques: a critical review. *J Cereb Blood Flow Metab* 20(3):423–451.

Lisman JE, Grace AA (2005) The hippocampal-VTA loop: controlling the entry of information into long-term memory. *Neuron* 46:703–713.

Litt MD, Cooney NL, Morse P (2000) Reactivity to alcohol-related stimuli in the laboratory and in the field: predictors of craving in treated alcoholics. *Addiction* 95:889–900.

Löber S, Croissant B, Heinz A, Mann K, Flor H (2006) Cue exposure in the treatment of alcohol dependence: effects on drinking outcome, craving and self-efficacy. *Br J Clin Psychol* 45(Pt 4):515–529.

Lodge DJ, Grace AA (2008) Augmented hippocampal drive of mesolimbic dopamine neurons: a mechanism of psychostimulant sensitization. *J Neurosci* 28:7876–7882.

Logothetis NK (2002) The neural basis of the blood-oxygen-level-dependent functional magnetic resonance imaging signal. *Philos Trans R Soc Lond B Biol Sci* 357(1424):1003–1037.

Ludwig AM, Wikler A (1974) Craving and relapse to drink. *Q J Stud Alcohol* 35:108–130.

Martinez D, Gil R, Slifstein M, Hwang DR, Huang Y, Perez A, Kegeles L, Talbot P, Evans S, Krystal J, Laruelle M, Abi-Dargham A (2005) Alcohol dependence is associated with blunted dopamine transmission in the ventral striatum. *Biol Psychiatry* 58:779–786.

Modell JG, Mountz JM (1995) Focal cerebral blood flow change during craving for alcohol measured by SPECT. *J Neuropsychiatry Clin Neurosci* 7:15–22.

Monti PM, Abrams DB, Binkoff JA, Zwick WR, Liepman MR, Nirenberg TD, Rohsenow DJ (1990) Communication skills training, communication skills training with family and cognitive behavioural mood management-training for alcoholics. *J Stud Alcohol* 51:263–270.

Myrick H, Anton RF, Li XB, Henderson S, Drobes D, Voronin K, George MS (2004) Differential brain activity in alcoholics and social drinkers to alcohol cues: relationship to craving. *Neuropsychopharmacology* 29:393–402.

Myrick H, Anton RF, Li X, Henderson S, Randall PK, Voronin K (2008) Effect of naltrexone and ondansetron on alcohol cue-induced activation of the ventral striatum in alcohol-dependent people. *Arch Gen Psychiatry* 65(4):466–475.

O'Brien CP (2005) Anticraving medications for relapse prevention: a possible new class of psychoactive medications. *Am J Psychiatry* 162:1423–1431.

Ogawa S, Lee TM, Nayak AS, and Glynn P (1990) Oxygenation-sensitive contrast in magnetic resonance image of rodent brain at high magnetic fields. *Magn Reson Med* 14:68–78.

Paulus MP, Tapert SF, Schuckit MA (2005) Neural activation patterns of methamphetamine-dependent subjects during decision making predict relapse. *Arch Gen Psychiatry* 62(7):761–768.

Ragozzino ME (2007) The contribution of the medial prefrontal cortex, orbitofrontal cortex, and dorsomedial striatum to behavioral flexibility. *Ann N Y Acad Sci* 1121:355–375.

Robbins TW, Everitt BJ (2002). Limbic-striatal memory systems and drug addiction. *Neurobiol Learn Mem* 78: 625–636.

Robinson TE, Berridge KC (1993) The neural basis of drug craving—an incentive-sensitization theory of addiction. *Brain Res Brain Res Rev* 18:247–291.

Rohsenow DJ, Monti PM, Rubonis AV, Sirota AD, Niaura RS, Colby SM, Wunschel SM, Abrams DB (1994) Cue reactivity as a predictor of drinking among male alcoholics. *J Consult Clin Psychol* 62:620–626.

Rossetti ZL, Melis F, Carboni S, Gessa GL (1992) Dramatic depletion of mesolimbic extracellular dopamine after withdrawal from morphine, alcohol or cocaine—a common neurochemical substrate for drug-dependence. *Ann N Y Acad Sci* 654:513–516.

Schneider F, Habel U, Wagner M, Franke P, Salloum JB, Shah NJ, Toni I, Sulzbach C, Hönig K, Maier W, Gaebel W, Zilles K (2001) Subcortical correlates of craving in recently abstinent alcoholic patients. *Am J Psychiatry* 158:1075–1083.

Schultz W, Dayan P, Montague PR (1997) A neural substrate of prediction and reward. *Science* 275:1593–1599.

Shalev U, Grimm JW, Shaham Y (2002) Neurobiology of relapse to heroin and cocaine seeking: a review. *Pharmacol Rev* 54:1–42.

Shalev U, Highfield D, Yap J, Shaham Y (2000) Stress and relapse to drug seeking in rats: studies on the generality of the effect. *Psychopharmacology* (Berl) 150(3):337–346.

Sharp PE (1999) Complimentary roles for hippocampal versus subicular/entorhinal place cells in coding place, context, and events. *Hippocampus* 9:432–443.

Siegel S (1975) Evidence from rats that morphinetolerance is a learned response. *J Comp Physiol Psychol* 89:498–506.

Siegel S (1999) Drug anticipation and drug addiction. The 1998 H. David Archibald Lecture. *Addiction* 94(8):1113–1124.

Siegel S, Hinson RE, Krank MD, McCully J (1982) Heroin overdose death—contribution of drug-associated environmental cues. *Science* 216(4544):436–437.

Srisurapanont M, Jarusuraisin N (2005) Naltrexone for the treatment of alcoholism: a meta-analysis of randomized controlled trials. *Int J Neuropsychopharmacol* 8:267–280.

Stark CE, Squire LR (2001) When zero is not zero: the problem of ambiguous baseline conditions in fMRI, *Proc Natl Acad Sci USA* 98:12760–12766.

Stewart J, Dewit H, Eikelboom R (1984) Role of unconditioned and conditioned drug effects in the self-administration of opiates and stimulants. *Psychol Rev* 91:251–268.

Streeton C, Whelan G (2001) Naltrexone, a relapse prevention maintenance treatment of alcohol dependence: a meta-analysis of randomized controlled trials. *Alcohol Alcohol* 36:544–552.

Tapert SF, Brown GG, Baratta MV, Brown SA (2004) FMRI BOLD response to alcohol stimuli in alcohol dependent young women. *Addict Behav* 29:33–50.

Tiffany ST (1990) A cognitive model of drug urges and drug-use behavior—role of automatic and nonautomatic processes. *Psychol Rev* 97:147–168.

Tobler PN, Fiorillo CD, Schultz W (2005) Adaptive coding of reward value by dopamine neurons. *Science* 307:1642–1645.

Tsai G, Gastfriend DR, Coyle JT (1995) The glutamatergic basis of human alcoholism. *Am J Psychiatry* 152(3):332–340.

Verheul R, van den Brink W, Geerlings P (1999) A three-pathway psychobiological model of craving for alcohol. *Alcohol Alcohol* 34(2):197–222.

Vezina P, Giovino AA, Wise RA, Stewart J (1989) Environment-specific cross-sensitization between the locomotor activating effects of morphine and amphetamine. *Pharmacol Biochem Behav* 32:581–584.

Volkow ND, Fowler JS, Wang GJ, Swanson JM (2004) Dopamine in drug abuse and addiction: results from imaging studies and treatment implications. *Mol Psychiatry* 9(6):557–569.

Volkow ND, Wang GJ, Fowler JS, Logan J, Hitzemann R, Ding YS, Pappas N, Shea C, Piscani K (1996) Decreases in dopamine receptors but not in dopamine transporters in alcoholics. *Alcohol Clin Exp Res* 20:1594–1598.

Volkow ND, Wang GJ, Telang F, Fowler JS, Logan J, Childress AR, Jayne M, Ma Y, Wong C (2006) Cocaine cues and dopamine in dorsal striatum: mechanism of craving in cocaine addiction. *J Neurosci* 26(24):6583–6588.

Volpicelli JR, Watson NT, King AC, Sherman CE, O'Brien CP (1995) Effect of naltrexone on alcohol "high" in alcoholics. *Am J Psychiatry* 152(4):613–615.

Vorel SR, Liu X, Hayes RJ, Spector JA, Gardner EL (2001) Relapse to cocaine-seeking after hippocampal theta burst stimulation. *Science* 292:1175–1178.

Weiss F (2005) Neurobiology of craving, conditioned reward and relapse. *Curr Opin Pharmacol* 5:9–19.

Weiss RD, Griffin ML, Mazurick C, Berkman B, Gastfriend DR, Frank A, Barber JP, Blaine J, Salloum I, Moras K (2003) The relationship between cocaine craving, psychosocial treatment, and subsequent cocaine use. *Am J Psychiatry* 160:1320–1325.

Wikler A (1948) Recent progress in research on the neurophysiologic basis of morphine addiction. *Am J Psychiatry* 105(5), 329–338.

Wilson SJ, Sayette MA, Fiez JA. (2004) Prefrontal responses to drug cues: a neurocognitive analysis. *Nat Neurosci* 7(3):211–214.

Wise RA (1988) The neurobiology of craving: implications for the understanding and treatment of addiction. *J Abnorm Psychol* 97(2):118–132.

Wong DF, Kuwabara H, Schretlin D, Bonson K, Zhou Y, Nandi A, Brasic J, Kimes AS, Maris MA, Kumar A, Contoreggi C, Links J, Ernst M, Rousset O, Zukin S, Grace AA, Rohde C, Jasinski DR, Gjedde A and London ED (2006) Increased occupancy of dopamine receptors in human striatum during cue-elicited cocaine craving. *Neuropsychopharmacology* 31:2716–2727.

World Health Organization (1992) *International statistical classification of diseases and related problems*, 10th ed. Geneva: World Health Organization.

Wrase J, Grüsser SM, Klein S, Diener C, Hermann D, Flor H, Mann K, Braus DF, Heinz A (2002) Development of alcohol-associated cues and cue-induced brain activation in alcoholics. *Eur Psychiatry* 17:287–291.

Wrase J, Schlagenhauf F, Kienast T, Wustenberg T, Bermpohl F, Kahnt T, Beck A, Strohle A, Juckel G, Knutson B, Heinz A (2007) Dysfunction of reward processing correlates with alcohol craving in detoxified alcoholics. *Neuroimage* 35(2):787–794.

5 Integrating Behavioral and Molecular Approaches in Mouse
Self-Administration Studies

Danielle L. Graham and David W. Self

CONTENTS

5.1 INTRODUCTION

Drug addiction is a serious mental illness involving severe motivational disturbances and a loss of behavioral control leading to personal devastation. The behavioral symptoms that accompany drug addiction can be modeled in rodents based on changes in their drug self-administration behavior. These symptoms include compulsive and escalating amounts of drug intake, along with a propensity for drug-seeking behavior in withdrawal, reflecting aberrations in the neural substrates that regulate these behaviors. Animal drug self-administration studies attempt to link neurobiological changes with the manifestation of specific behavioral symptoms. Since most changes in neuronal function stem from molecular "neuroadaptations" that occur in specific cell types in anatomically discrete brain regions, modern technological advances can be used to manipulate single gene targets with similar anatomical precision, and at postdevelopmental stages of adulthood, in order to mimic the neuroadaptations produced by chronic drug use. This approach is necessary to delineate important functional interactions that underlie the etiology of primary disease symptoms.

In this chapter, we describe recent drug self-administration studies that implement modern molecular genetic approaches to manipulate target proteins in mice.

We begin by describing behavioral approaches for studying drug-taking and drug-seeking behaviors using the self-administration model, and the important influence of genetic background on self-administration behavior. A serial behavioral testing procedure is described for determination of operant learning capacity with natural rewards, followed by analysis of intravenous drug self-administration using different reinforcement schedules to assess drug-taking and drug-seeking behaviors. The integration of these procedures with inducible transgenic and localized knockout approaches to study gain and loss of protein function is highlighted in mouse self-administration studies.

A high degree of temporal and anatomical control over *in vivo* target protein expression is required for meaningful results in behavioral/systems-level neuroscience, and it is usually desirable to limit effects to postdevelopmental or adult stages. Thus, while constitutive and diffuse transgenic expression or genetic deletion can provide useful information in certain circumstances, they rarely provide definitive information on the contribution of drug-induced neuroadaptations to the expression of addictive behavior, since these neuroadaptations generally occur with a high degree of regional and cellular specificity. More recent advances in molecular and genetic technology allow for better control over cell-specific expression and localized genetic deletion, a preferable approach to bolster the significance of relevant but labor-intensive drug self-administration studies in mice.

5.2 DISSOCIATION OF DRUG-TAKING AND DRUG-SEEKING BEHAVIORS USING SELF-ADMINISTRATION MODELS

Most mammalian species, including rodents and nonhuman primates, will learn to self-administer many of the same drugs that are abused by humans, reflecting drug actions on evolutionarily conserved natural reinforcement substrates (Deneau et al. 1969; Schuster and Thompson 1969; Collins et al. 1984; Johanson and Fischman 1989). In humans, initial acquisition of drug self-administration, or recreational drug use, often develops into a state of addiction characterized by increasing amounts of drug consumption during a self-administration binge, and by episodes of intense drug craving during abstinence that can trigger relapse to drug-seeking behavior (American Psychiatric Association 2000). In rodent self-administration experiments, the transition from premorbid drug use to addicted biological states is modeled by prolonging daily access to the drug, which induces an escalation in the rate of drug intake (Ahmed and Koob 1998; Deroche et al. 1999), or selected from outbred rat populations based on higher individual differences in preferred levels of drug intake (Piazza et al. 2000; Sutton et al. 2000; Deroche-Gamonet 2004; Edwards et al. 2007). Animals with higher drug intake also show an enhanced propensity for drug-seeking behavior during periods of forced abstinence, thereby encompassing both addictive traits in these animal models.

The schedule of drug reinforcement is an important consideration when modeling distinct symptoms of addictive behavior in self-administration studies. When drug self-administration is not restricted by high response requirements or by prolonged intervals of drug unavailability, animals titrate their preferred level of drug intake in a highly stable manner. For example, when animals are required to perform a fixed low number of lever presses, such as a fixed ratio (FR) of five lever presses to receive each drug injection, animals compensate for a lowering in the injection dose by increasing the rate of self-administration. Thus, self-administration rates on FR reinforcement schedules are remarkably dose-dependent and are inversely related to the dose received with each self-injection. High injection doses with prolonged effects reduce the rate of self-administration, whereas moderate injection doses with brief effects increase self-administration rates. This relationship is reflected by an inverted U-shaped self-administration dose-response curve, spanning subthreshold doses that are too low to support self-administration, moderate suprathreshold doses that are self-administered at increased rates, and higher doses that are self-administered at reduced rates due to prolonged effects of the drug injections. An addicted biological state is reflected by increases in the rate of drug intake at a given injection dose, and such compensatory behavior suggests that the perceived pharmacological impact of the drug injections is reduced in addicted animals. The state of addiction also is associated with a vertical shift in the inverted U-shaped dose-response function, reflecting escalating drug intake at suprathreshold doses on less demanding FR reinforcement schedules (Ahmed and Koob 1998; Piazza et al. 2000). Thus, the contribution of molecular neuroadaptations to drug addiction ultimately must determine whether these biological changes contribute to escalating drug intake and a vertical shift in the FR dose-response curve.

In contrast, drug-*seeking* behavior is measured under conditions when drug reinforcement is withheld, and it is thought to reflect behavior that would underlie craving and relapse during abstinence (Stewart et al. 1984; Robinson and Berridge 1993; Berridge and Robinson 1998; Wise 2004). Drug seeking is reflected by approach behavior aimed at performing responses that previously delivered drug injections. Most studies measure the level of effort (lever pressing) an animal will exert to obtain drug as an index of drug-seeking behavior, rather than the actual amount of drug consumed. There are numerous methods to measure drug-seeking behavior, but they can be divided into two general categories: drug seeking during active self-administration, and drug seeking during periods of forced abstinence. In progressive ratio (PR) testing, the response demands for each successive drug injection increase progressively during active self-administration, and the highest ratio of lever presses/injection achieved before animals quit responding is an index of a drug's reinforcing efficacy. The addicted phenotype is associated with profound increases in the pursuit of drug reinforcement on PR schedules (e.g., Piazza et al. 2000). However, certain neurobiological manipulations that increase drug seeking on PR schedules do not always produce escalating drug intake on less demanding FR schedules, indicating separate neurobiological regulation of drug-taking and drug-seeking behaviors, as discussed later. The use of both FR and PR schedules of drug reinforcement is a powerful combination for clarifying the contribution of specific neuroadaptations to escalating drug intake and the motivation for drugs when reinforcement is withheld.

Another key component characterizing the state of addiction is the propensity or vulnerability to relapse during abstinence from chronic drug self-administration.

The propensity for relapse during abstinence can be modeled in rodents using the extinction/reinstatement paradigm. This model of drug craving and relapse has face validity because environmental and pharmacological stimuli that elicit relapse to drug seeking in animals also trigger drug craving in humans (Jaffe et al. 1989; Robbins et al. 1997; Sinha et al. 1999). In the extinction phase of this procedure, drug-seeking behavior is measured by the magnitude and persistence of nonreinforced responding at the lever that delivered drug injections (drug-paired lever) during prior self-administration. Drug seeking in extinction is elicited by environmental and contextual cues associated with drug use in the self-administration test chambers and inevitably diminishes with repeated training in the absence of drug reinforcement. Following extinction of drug-seeking behavior, the ability of specific experimenter-delivered stimuli to elicit or "reinstate" drug-paired lever responding is measured. In rats, reinstatement of drug-seeking behavior can be induced by priming injections of drugs, by presentation of discrete cues associated with drug injections, and by brief exposure to moderate intermittent footshock stress.

While there are numerous extinction/reinstatement studies in rats, there are very few reports of successful reinstatement of drug seeking in mice. Thus, for example, the C57BL/6J strain of mice avidly self-administer intravenous cocaine or methamphetamine, but this strain does not exhibit effective reinstatement of drug seeking following intraperitoneal (ip) priming injections of cocaine or methamphetamine (Fuchs et al. 2003; Yan et al. 2006). However, drug seeking can be elicited in C57BL/6J mice by contextual drug-associated stimuli during extinction testing, or by response-contingent administration of discrete injection cues in reinstatement testing (Fuchs et al. 2003; Yan et al. 2006). Studies show that extinguished C57Bl/6J mice will reinitiate self-administration behavior when responding is reinforced by intravenous (iv) cocaine or methamphetamine injections (Yan et al. 2006; Kruzich 2007), but this behavior reflects a return to drug taking reinforced by drug injections rather than an accurate measure of drug seeking in the absence of reinforcement. Since the C57Bl/6J strain is a common genetic background for knockout and transgenic manipulations, this caveat has limited most mouse models of addiction to studies on self-administration rather than relapse behaviors. Finally, other strains such as 129X1/SvJ exhibit modest reinstatement of cocaine seeking albeit at a very high dose (6.0 mg/kg iv; Highfield et al. 2002), and we have observed effective dose-dependent (3–10 mg/kg ip) reinstatement of cocaine seeking in a mixed background of 25% SJL, 25% C57BL/6J and 50% ICR (Institute for Cancer Research) (D.W. Self, unpublished observations). These latter findings suggest that more work is needed to establish appropriate genetic backgrounds for studies on relapse using drug-primed reinstatement.

5.3 INFLUENCE OF GENETIC BACKGROUND ON DRUG SELF-ADMINISTRATION

Unlike commonly used outbred rat strains, most mice are maintained as inbred strains with substantial within-strain homogeneity but substantial heterogeneity between mouse strains. These genetic backgrounds markedly differ in their behavioral responses to drugs of abuse. Thus, the choice of genetic background for mouse self-administration studies is critical. The C57BL/6J mouse strain is one of the most commonly used inbred strains for the generation and maintenance of knockout and transgenic mice (Sedivy and Joyner 1992; Spanagel and Sanchis-Segura 2003). The behavioral response to several drugs of abuse differs substantially between C57BL/6J and many other inbred mouse strains. For example, C57BL/6J mice show an intermediate propensity to acquire cocaine self-administration behavior, showing somewhat lower acquisition rates than DBA/2J mice, but considerably greater acquisition than 129 and BALB/c strains. However, peak self-administration rates and cocaine intake are highest in C57BL/6J mice across multiple injection doses, suggesting that C57BL/6J mice display an inherently cocaine-addicted phenotype compared to other strains after initial acquisition of self-administration (Morse et al. 1993; Deroche et al. 1997; Rocha et al. 1998b; Kuzmin and Johansson 2000). C57BL/6J mice readily acquire and maintain cocaine self-administration using a nose-poke operant response when experimentally naïve, or following lever-press training with sucrose reinforcement (Deroche et al. 1997; Ruiz-Durantez et al. 2006). Several other studies have found reliable oral or intravenous cocaine self-administration with C57Bl/6J mice (Carney et al. 1991; Grahame and Cunningham 1995; Grahame et al. 1995). Although the C57BL/6J strain demonstrates robust cocaine self-administration in a variety of paradigms, this strain may be less than optimal for detecting the addiction-promoting actions of genetic manipulation since their self-administration behavior may already reflect the upper limits of an addicted phenotype (ceiling effect).

We tested the ability of two common mouse strains to acquire intravenous cocaine and heroin self-administration under identical conditions in the same laboratory. In these experiments, all mice underwent prior lever-press training with food pellets as a reinforcer. This procedure allows experiments to bypass potential strain differences in operant learning capacity and to focus on initial sensitivity to drug reinforcement using a prelearned task. As shown in Figure 5.1A, both C57BL/6J and BALB/c strains initiate cocaine self-administration at similar rates on the first test day, reflecting their prior lever-press training. However, BALB/c mice fail to acquire and maintain cocaine self-administration using a moderate training dose, and rapidly extinguish virtually all self-administration behavior by the second test day. This rapid and complete extinction of self-administration behavior suggests that BALB/c actually avoid cocaine injections rather than lack sensitivity to cocaine. BALB/c mice may be more sensitive to the anxiogenic properties of cocaine, since pretreatment with diazepam enhances acquisition of cocaine self-administration in this strain (David et al. 2001a). In contrast, C57BL/6J mice reliably acquire cocaine self-administration at this dose (Figure 5.1A), consistent with numerous studies described above. Conversely, Figure 5.1B shows that heroin is self-administered at higher levels in BALB/c mice compared to C57BL/6J mice using a suprathreshold training dose (15 µg/kg/injection), suggesting that BALB/c and C57BL/6J strains exhibit differential preferences for cocaine and heroin. These data are supported by other studies indicating that experimentally naïve BALB/c mice will not acquire cocaine self-administration when cocaine is administered alone via nose-poke responses but will self-administer cocaine in combination with heroin as a "speedball" cocktail (Deroche et al. 1997; David et al. 2001b). Together, these findings suggest that BALB/c strain is inappropriate for molecular genetic studies of cocaine

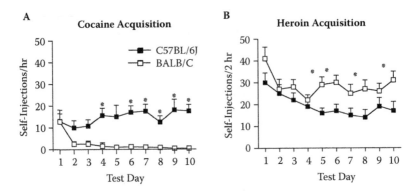

FIGURE 5.1 Inbred strain differences in acquisition of intravenous cocaine and heroin self-administration. (A) The C57BL/6J mice acquire cocaine self-administration (20 μg/injection or ~600 μg/kg/injection) while the BALB/c mice rapidly extinguish cocaine self-administration at this moderate-injection dose, despite prior lever-press training for food pellets. * $P < 0.05$ indicates C57/BL6 mice self-administered significantly more cocaine than BALB/c mice. (B) Both C57BL/6J and BALB/c strains acquire heroin self-administration (15 μg/kg/injection), although BALB/c self-administer significantly more heroin than C57BL/6J mice. * $P < 0.05$ indicates BALB/c mice self-administered heroin significantly more than C57BL/6J mice. Data are expressed as mean ± SEM, $N = 9–11$/group.

addiction unless a better understanding of cocaine-induced anxiety is desired. In contrast, the C57BL/6J strain may be better suited for analyzing transgenic expression in studies on heroin addiction than BALB/c, since the C57BL/6J background displays intermediate sensitivity to heroin reinforcement potentially amenable to addiction-promoting effects of genetic manipulation.

Another potential caveat to using inbred mouse strains is that isogenic backgrounds are highly susceptible to within-strain epistatic interactions that mask or accentuate the effects of single-gene manipulations (Palmer et al. 2003). Epistasis is due to interactions with specific native genes that alter the biological or behavioral phenotype of the targeted genetic manipulation. The potential for epistasis is higher with inbred mouse strains that are relatively homogeneous in potential epistatic interactions across individual mice. Thus, it is useful to compare the behavioral effects of genetic manipulation in multiple unrelated inbred strains, each with a different complement of potential epistatic interactions, to clearly establish that the observed behavioral change is caused by the manipulated target gene and is not peculiar to strain-specific epistatic interactions. This approach is cumbersome and not feasible for labor-intensive mouse self-administration studies.

An alternative approach involves using outbred mice strains with substantial within-strain heterogeneity in potential epistatic interactions. Outbred strains also generally exhibit greater fertility and viability compared to inbred strains such as C57BL/6J, a major advantage when maintaining mouse lines for investigator use. However, within-strain heterogeneity also can lead to highly variable behavioral effects. For example, we studied acquisition of cocaine self-administration in outbred ICR mice and found that a relatively low percentage (< 20%) of these outbred mice acquire intravenous cocaine self-administration. Thus, this nonpreferring genetic background was optimized for cocaine self-administration studies by incrementally adding increasing percentages of C57BL/6J genes into the outbred ICR background (Ruiz-Durantez et al. 2006).

To generate mixed offspring, inbred C57BL/6J mice were crossed with outbred ICR mice to produce F1 offspring that were 50% C57BL/6J: 50% ICR. Several different ICR parents were used to control for potential founder effects in the outbred ICR lineage. Some of the resulting F1 generation (50:50) were backcrossed to pure parental C57BL/6J mice to produce a second generation of 75% C57BL/6J: 25% ICR mice. Similarly, other F1 mice were backcrossed to pure parental ICR mice to produce 25% C57BL/6J: 75% ICR mice. The F1 and F1 x parental hybrids were compared with the pure parental strains for acquisition of cocaine self-administration.

While all mice in parental and hybrid strains readily acquire sucrose self-administration (Ruiz-Durantez et al. 2006), Figure 5.2A shows that only mice with 75% and 100% C57BL/6J genes acquire cocaine self-administration. In contrast, mice with ≤50% C57BL/6J genes fail to acquire cocaine self-administration. The mean self-administration rates for each parental strain or hybrid reflect the percentage of mice that meet acquisition criteria, defined as ≥15 cocaine injections on the last 3 days of acquisition testing. Thus, while only 19% of pure ICR mice acquire cocaine self-administration, the presence of 50%, 75%, and 100% C57BL/6J genes progressively increases the percentage the population meeting acquisition criteria to 31%, 52%, and 75%, respectively (Figure 5.2A). Moreover, mice with ≥75% C57BL/6J genes that ultimately acquire cocaine self-administration do so faster than the few mice with ≤50% C57BL/6J genes that acquire self-administration. These data are consistent with a genetically based dose-dependent enhancement of cocaine reinforcement by C57BL/6J genes in the nonpreferring ICR background. Presumably, increasing percentages of C57BL/6J genes increases the likelihood that critical reinforcement-enhancing genes are present.

We subjected mice that acquire cocaine self-administration to dose-response testing on an FR1 reinforcement schedule. Figure 5.2B shows that hybrids containing 75% C57BL/6J and 25% ICR genes self-administer cocaine with a typical inverted U-shaped dose-response curve, whereas the dose-response curve for mice with ≤50% C57BL/6J genes is flat and generally shifted downward. While mice in both groups that acquire cocaine self-administration do so with very regular temporal patterns of drug intake (Ruiz-Durantez et al. 2006), the failure of mice with ≤50% C57BL/6J genes (≥50% ICR) to show an inverted U-shaped dose-response curve may reflect relatively less sensitivity to cocaine reinforcement. Indeed, lower peak self-administration rates in mice with ≤50% C57BL/6J genes suggest that cocaine is less reinforcing than in mice with ≥75% C57BL/6J genes. Interestingly, pure C57BL/6J mice lacking ICR genes fail to exhibit compensatory reductions in self-administration rates with increases in the unit dose/injection (Figure 5.2B). This effect is due to a dose-dependent increase in stereotypic burst responding for cocaine in C57BL/6J mice under the FR1 reinforcement schedule (Ruiz-Durantez et al. 2006). However, C57BL/6J mice will exhibit an inverted U-shaped dose-response curve for cocaine self-administration under higher FR response requirements (e.g., FR5) or with prolonged training at each dose (D.W. Self, unpublished observations). Another recent study used a prolonged postinjection time-out period and a 30 mg/kg cap on total

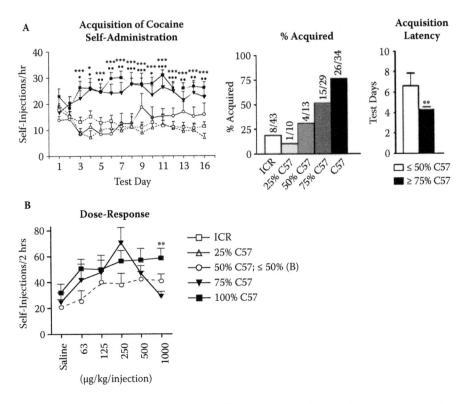

FIGURE 5.2 (A) Contribution of C57BL/6J genes to acquisition of intravenous cocaine self-administration when crossed with mice with a nonpreferring genetic background (outbred ICR). Increasing the percentage of C57BL/6J genes in outbred ICR mice facilitates the acquisition of cocaine self-administration reflected by the number of self-injections per hour (left), the percentage of mice achieving acquisition criteria (middle), and the mean latency to acquire cocaine self-administration in pooled groups of ≤50 or ≥75% C57BL/6J genes in mice that met acquisition criteria (right). * $P < 0.05$, ** $P < 0.01$, and *** $P < .001$ compared with the ICR strain (or ≤50% C57BL/6J). (B) Inverted U-shaped self-administration dose-response curve for mice containing 75% C57BL/6J genes ($n = 14$), but not in mice with ≤50% C57BL/6J genes ($n = 8$) or with 100% C57BL/6J genes ($n = 13$). Mice with 100% C57BL/6J genes self-administered more cocaine at a higher test dose (1000 μg/kg/injection) than mice with 75% C57BL/6J genes, reflecting their inability to respond to an increase in injection dose with decrease in self-administration rates. ** $P < 0.01$ compared with 75% C57BL/6J mice. (From Ruiz-Durantez et al., *Psychopharmacology* 186, 553, 2006. With permission.)

cocaine intake to facilitate inverted U-shaped dose-response curves in C57BL/6J mice on an FR1 reinforcement schedule (Thomsen and Caine 2006). In this regard, the presence of 25% ICR genes in C57Bl/6J mice could represent a useful approach for rapidly obtaining dose-response sensitivity when access to cocaine is less restricted (FR1).

These findings suggest that heritable traits impart a substantial genetic load that facilitates the propensity for cocaine addiction among individuals in outbred populations. Furthermore, the 75% C57BL/6J × 25% ICR hybrid may provide an optimal genetic background for the expression of targeted genetic manipulations in cocaine self-administration studies, since an intermediate behavioral phenotype would be sensitive to both addiction-promoting and addiction-opposing influences. Thus, about 50% of these mice acquire cocaine self-administration at suprathreshold doses, and so the effects of targeted genetic manipulation on the propensity of the "population" to acquire self-administration can be determined. In contrast, a more common approach determines the threshold injection dose that is necessary for acquisition, reflecting pharmacological sensitivity. However, low-dose sensitivity arguably is less directly related to cocaine-addicted phenotypes (Ahmed and Koob 1998) than a propensity for acquisition at suprathreshold injection doses. In addition, the presence of even 25% outbred ICR genes in 75% C57BL/6J offspring increases mean litter size and survivability compared to pure C57BL/6J mice, particularly when the maternal parent carries ICR genes (Ruiz-Durantez et al. 2006).

5.4 CELL-SPECIFIC TRANSGENIC EXPRESSION IN DRUG SELF-ADMINISTRATION STUDIES

Transgenic overexpression is used to study gain of target protein function. Transgenic mice are generated via insertion of new genetic material into the host strain's genomic DNA (Carter 2004; Wells and Carter 2001). A transgene is synthesized *in vitro*, injected into the nucleus of a fertilized egg, and then implanted into a pseudopregnant female for gestation. The transgene is incorporated into the zygotic genome through recombination during cell division. While all cells contain a copy of the transgene, studies employ neuron-specific promoters placed upstream of the target gene to limit expression to the central nervous system. For example, the promoter region of the neuron-specific enolase gene is often used to drive transgenic expression, since it will only be activated in neurons. Ultimately, the regional and cell-specific pattern of gene expression in the brain also involves regulatory interactions with host genetic sequences surrounding a randomly determined site of transgenic insertion, leading to a variety of expression patterns across multiple lines expressing the same transgene. The role of increased target protein expression in drug self-administration behavior can depend on the particular cell type or brain region where it is expressed. Specific hypothetical interactions are tested by selecting a transgenic mouse line with the desired expression profile from multiple lines with different or overlapping expression patterns.

These randomly generated expression profiles are based on the site of transgenic insertion in the host genome, while targeted expression incorporates specific promoters for genes with known regional and cell-specific expression patterns in the brain. For example, the expression of calcium-calmodulin kinase II is naturally limited to the forebrain, and incorporating the promoter for calcium-calmodulin kinase II in a transgene can produce a similar region-specific transgenic expression profile. Similarly, certain cell types such as norepinephrine-producing cells contain cell-specific proteins (dopamine β-hydroxylase), and promoters for dopamine β-hydroxylase will limit target protein expression to norepinephrine-containing cells (Wells and Carter 2001; Morozov et al. 2003). Another approach that circumvents the need for prior knowledge of specific promoter sequences instead uses very large DNA sequences containing numerous and unspecified promoter regions flanking a cell-specific protein marker. These promoter-containing sequences will naturally limit transgenic expression to neuronal subpopulations expressing the protein marker. This approach has been used to target transgenic expression to GABAergic interneurons or D1- and D2-receptor-containing cells (Gong et al. 2003). While these cell-targeted approaches in some cases may lack the regional selectivity needed for systems/behavioral work *in vivo*, future technological developments for both cell- and region-specific targeting promise to allow an increasing level of anatomical control over transgenic expression patterns.

A major advantage of transgenic overexpression is the absolute specificity of protein modulation when compared to pharmacological approaches. However, a prominent disadvantage is the common incidence of developmental or homeostatic compensation that can occur even with relatively brief periods of continuous overexpression (Self 2005). Inducible transgenic systems allow the onset of transgenic expression to occur in adult stages that circumvent developmental compensation. While homeostatic compensatory changes may still ensue, inducible but continuous overexpression is highly suited for drug addiction studies to mimic the enduring molecular changes produced by chronic drug self-administration.

Tetracycline-inducible transgenic expression involves the bitransgenic expression of a target protein regulated by a Tet-Op promoter and a second transgene for the tetracycline transactivator protein (tTA) that is expressed continuously (Morozov et al. 2003). The tTA protein binds the Tet-Op promoter region of the target protein transgene to influence its expression. The addition of low-dose tetracycline, or more commonly, doxycycline, to the drinking water binds and inactivates the tTA protein, preventing its interaction with the Tet-Op promoter of the primary transgene (Figure 5.3). Both Tet-on and Tet-off systems can be constructed based on whether the Tet-Op promoter is situated to drive or inhibit transgene expression. Potential drawbacks to transgenic overexpression systems include diffuse expression in relatively few cells that may produce only marginal behavioral effects, leaks in expression beyond Tet-Op regulation, and other cellular processes including trafficking, degradation, and saturating levels of endogenous regulatory proteins that limit the efficacy of overexpression. The anatomical profile of transgenic expression in inducible bigenic systems is related to overlapping intracellular expression of both tTA and Tet-Op-driven transgenes, but other factors that are poorly understood play a role in determining observed expression patterns. Nevertheless, the availability of multiple tTA driver mouse lines with well characterized region- and cell-specific expression profiles would be an enormous asset for targeting transgenic expression in specific cell types and anatomically discrete brain regions in order to mimic the pattern of protein regulation that occurs in addicted biological states.

We used the Tet-inducible transgenic system to study the effects of the transcription factor ΔFosB on cocaine self-administration behavior. Chronic exposure to cocaine induces ΔFosB exclusively in dynorphin-containing striatal neurons that constitute the "direct" striatal output pathway (Nye et al. 1995; Moratalla et al. 1996). Using a bigenic tetracycline-inducible system described above, transgenic

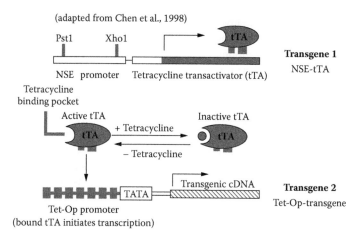

FIGURE 5.3 Inducible transgenic approach for expression of a transgene via the Tet-Op bigenic strategy. Transgenic mice containing the gene for the tetracycline transactivator (tTA) under the direction of a neuron-specific enolase (NSE) promoter are crossed with mice containing the transgene of the target protein under the direction of the Tet-Op promoter. When doxycycline is present in the drinking water, tTA is inactivated and expression of the transgene is suppressed. Removal of doxycycline activates tTA to induce expression of the transgene. Different lines of tTA mice convey neuronal-specific expression due to the NSE promoter, and regional or cell type-specific expression based on cellular regulatory control of genomic loci where the tTA and target protein transgenes are inserted. From Self, D. W., Comparison of transgenic strategies for behavioral neuroscience studies in rodents, *Psychopharmacology*, 147:35–37, 1999. With permission.

mice were generated that overexpress ΔFosB specifically in dynorphin-containing striatal neurons to study the role of this drug-induced neuroadaptation in cocaine addiction (Kelz et al. 1999). These mice, along with their bigenic littermate controls maintained on doxycycline to suppress the transgene, were tested in a serial behavioral testing procedure with both natural and drug rewards (Colby et al. 2003). The rationale and methodology for this serial testing procedure is described in detail below using ΔFosB mice to exemplify the approach.

It is important to consider that genetic manipulation can often alter spontaneous locomotor activity and exploratory behavior that would indirectly influence acquisition of operant tasks independent of reinforcement mechanisms. Other genes are involved in the formation of stimulus–response associations that underlie operant learning itself. However, the role of these initial learning processes in the transition from recreational drug use to addicted biological states is unclear, since they occur whether or not drug self-administration ultimately leads to the development of addiction. Thus, studies on the neurobiological mechanisms of operant learning may address a fundamentally different question than studies aimed at understanding the biological basis of drug addiction. Moreover, generalized alterations in performance can inadvertently influence several postacquisition measures of addictive behavior, including self-administration on FR and PR reinforcement schedules, and nonreinforced responding in extinction/reinstatement studies. For example, many genetic manipulations produce a generalized perseveration of learned behavioral responses, and these effects inadvertently lead to artifactual enhancement of drug-seeking behavior when reinforcement is withheld on PR self-administration schedules or under extinction conditions. Such generalized response perseveration may not accurately reflect a role for protein overexpression in drug-seeking behavior.

In order to control for such generalized changes in performance and operant learning capacity in drug self-administration studies, transgenic and knockout mice often are subjected to operant testing with natural rewards such as food or sucrose pellets. Given the difficulty and expense of breeding and maintaining transgenic and knockout mouse lines, we developed a serial behavioral testing procedure to study operant learning capacity with natural rewards, followed by analysis of multiple drug self-administration behaviors in the same mice. The procedure determines acquisition and extinction rates of food-reinforced behavior, the dose threshold for acquisition of intravenous drug self-administration, followed by postacquisition dose-response analysis using FR and PR reinforcement schedules. The entire procedure can be conducted in about 11–13 weeks, depending on the time required for mice to meet criteria for specific behavioral tasks.

Tet-inducible transgenic mice are subjected to this procedure at least 8 weeks after removal of doxycycline from drinking water to allow calcium-bound doxycycline to clear and permit tTA to activate transgenic expression. As depicted in Figure 5.4, animals initially are maintained on a food-restricted diet and exposed to the operant chambers in the absence of reinforcement to compare spontaneous lever-press behavior in genetically altered and control mice. If spontaneous lever-press behavior substantially differs between transgenic and control mice, this behavior is habituated to avoid activity-related differences in subsequent acquisition of operant behavior. An example of spontaneous responding in inducible transgenic mice overexpressing the transcription factor ΔFosB is shown in Figure 5.5B. Both ΔFosB mice and their littermate controls maintained on doxycycline sample response levers at equivalent rates in the absence of reinforcement in a 1-hr test for spontaneous lever-press behavior. Since these mice show no difference in spontaneous lever-press behavior, further habituation of spontaneous responding is not required.

Acquisition of operant behavior reinforced by food or other natural rewards provides a rapid and reliable method for analyzing operant learning capacity independent of drug reinforcement, and can be used to facilitate subsequent acquisition of drug self-administration as shown in Figure 5.1. It is important to consider the background strain to determine the appropriate schedule of sucrose reinforcement. Different strains of mice will acquire lever-press behavior at different rates, and prolonging the latency to acquire lever-press behavior over several test sessions enhances the ability to detect differences in operant learning capacity. For ΔFosB expressing mice maintained on a mixed background (50% ICR, 25% C57BL/6J, 25% SJL), an FR3 schedule of food pellet reinforcement (three presses/pellet) leads to an average acquisition latency of ~5–6 test sessions (Figure 5.5A). In contrast, virtually 100% of pure C57BL/6J mice will acquire lever-press behavior by the second test session on the FR3 schedule, even when spontaneous lever-press behavior is habituated to ≤10 presses/hr prior to acquisition testing (Graham et al. 2007; D.W.

I. Determination of operant learning capacity

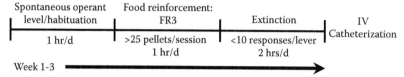

II. Analysis of drug self-administration behavior

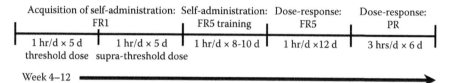

FIGURE 5.4 Schematic of time course for serial behavioral testing procedures used to assess drug self-administration behavior in transgenic or knockout mice. Prior to intravenous catheterization, operant behavior is analyzed to investigate potential activity- or learning-related influences of genetic manipulation unrelated to drug sensitivity or preference. The initial test involves spontaneous operant responding (lever press) in the absence of reinforcement, an important activity-related measure that could contribute to artifactual differences in acquisition rates. Differences in spontaneous responding can be habituated to low levels (<10 responses/lever) prior to acquisition of food or sucrose pellet self-administration where a fixed ratio (FR) reinforcement schedule of three lever presses/pellet is used. Depending on the genetic background strain, the ratio requirement (e.g., FR3) can be increased to prolong the number of test sessions until acquisition criteria are met, thereby enhancing sensitivity to potential differences in operant learning capacity. Following acquisition, lever-press behavior is extinguished over several test sessions in the absence of food reinforcement, and the latency to achieve extinction criteria provides an index of extinction learning. Genetic manipulations that produce response perseveration in extinction tests could confound certain measures of drug-seeking behavior in subsequent self-administration tests. Once extinction criteria are met, mice are surgically implanted with a chronic indwelling intrajugular catheter in a minor surgical procedure and allowed 3–4 days to recover. Dose thresholds for supporting acquisition of intravenous (IV) drug self-administration are tested in separate groups in the initial week of self-administration testing (FR1), followed by additional acquisition training using a suprathreshold dose to engender self-administration in all test mice. After acquisition, the response requirement is gradually raised to FR5, and stabilized animals are subjected to dose-response determinations on fixed (FR) and progressive (PR) ratio reinforcement schedules (2 days at each dose).

Self, unpublished observations). Thus, a higher FR response requirement may be necessary to prolong acquisition in this or other mouse strains. ΔFosB-expressing mice and their littermate controls acquire lever-press responding for food pellets at similar rates (Figure 5.5), indicating that ΔFosB does not alter operant learning capacity. Following acquisition and two additional days of lever-press training, ΔFosB mice and controls show similar lever discrimination biased toward the lever producing reinforcement, indicating similar abilities to inhibit generalized responding at the inactive lever. Both groups also exhibit similar latencies to consume the entire 30-pellet allotment (Figure 5.5B), an important control to ensure equivalent

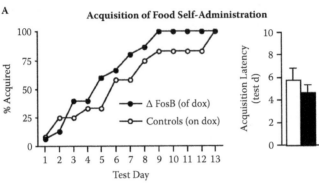

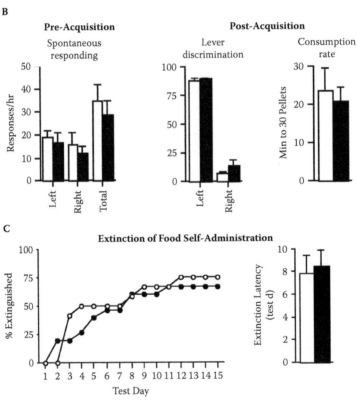

FIGURE 5.5 Analysis of operant learning capacity in mice with inducible striatal cell type–specific transgenic overexpression of ΔFosB. (A) Overexpression of ΔFosB (filled circles and bars) fails to alter acquisition of lever-press behavior reinforced by food pellets (FR3), as both ΔFosB-expressing mice and littermate controls (maintained on doxycycline) achieve acquisition criteria with similar mean latencies of ~5–6 days (# sessions until ≥25/session pellets are earned). (B) Prior to acquisition testing, spontaneous lever-press behavior in naïve mice is similar between groups (left), and ΔFosB overexpression fails to alter lever discrimination (middle), or the latency to consume the 30 food pellet allotment after 2 additional training days (right). (C) The mean number of test sessions required for achieving extinction criteria in the absence of food reinforcement (≤10 responses/lever in 2 hrs) is similar in ΔFosB and control mice. (From Colby et al., *J. Neurosci.*, 23, 2488, 2003. With permission.)

motivational states (hunger) between study groups that could negate observed differences in the rate of operant learning.

In some cases, prior lever-press training is desired to bypass potential transgenic deficits in operant responding and engender adequate and equivalent initial sampling of drug injections. This procedure will determine whether animals avoid drug self-administration (e.g., Figures 5.1 and 5.2). However, a major caveat to prior lever-press training with natural rewards is that response carryover can obscure potential differences in the rate of acquisition of drug self-administration in subsequent tests. One approach to circumvent this problem involves using alternate levers for food and drug reinforcement, although response generalization from one lever to another can remain a factor. Thus, in certain instances it is important to extinguish lever-press behavior before testing acquisition of drug self-administration behavior. Analysis of extinction from natural rewards also provides an important behavioral control for potential transgenic effects on extinction learning, or response perseveration, that would negate motivation-related differences in drug-seeking behavior as described above. The serial testing procedures depicted in Figure 5.4 incorporate an extinction phase to reduce lever-press behavior to low levels prior to surgical catheterization and drug self-administration testing. When mice overexpressing ΔFosB are subjected to extinction training in the absence of food pellets, they extinguish lever-press behavior at rates similar to controls, averaging ~8 test sessions to achieve an extinction criteria of ≤10 responses/lever in 2 hrs (Figure 5.5C). The data indicate that ΔFosB mice do not exhibit undue response perseveration or deficits in extinction learning when reinforcement is withheld. Certain mouse strains such as C57BL/6J are notably resistant to extinction of food-reinforced behavior (D.W. Self, unpublished observations), another reason to consider mixed genetic backgrounds for self-administration studies.

Following acquisition and extinction of food-reinforced responding, mice undergo surgical insertion of a chronic intrajugular catheter to examine intravenous drug self-administration behavior. While drugs of abuse are self-administered by several different routes, intravenous self-administration provides rapid temporal contingency between the behavioral response and the onset of drug action, and closely approximates the pharmacokinetics of the most addictive routes of human drug use (intravenous or inhalation). Intravenous self-administration also avoids confounding factors such as drug taste or excessive delays between drug delivery and subjective effects that accompany oral or intranasal routes of administration. Beginning 4–7 days after catheter implantation, dose thresholds are determined for acquisition of drug self-administration in daily 1-hr sessions over 5 consecutive days of testing (Figure 5.4). To facilitate this procedure, two doses are chosen that span a subthreshold and threshold dose needed to support acquisition of self-administration in control mice. This approach is designed to detect an enhancement in drug sensitivity at the subthreshold dose with genetic manipulation, while a suprathreshold dose would be necessary to detect a reduction in drug sensitivity; that is, a genetic manipulation would attenuate acquisition at the threshold dose, but not at a suprathreshold dose. Determination of dose thresholds is a sensitive measure of subtle changes in sensitivity that may ensue with anatomically restricted single-gene manipulation, while brainwide genetic manipulation may produce stronger effects on cocaine preference at suprathreshold doses, similar to the effects of genetic background shown in Figures 5.1 and 5.2. Thus, an important limitation of threshold dose determinations is that mice ultimately demonstrate a capacity to acquire drug self-administration at higher doses.

Using this procedure, inducible transgenic mice overexpressing ΔFosB acquire cocaine self-administration at a subthreshold dose of 125 μg/kg/injection, a dose that fails to support stable acquisition in littermate controls maintained on doxycycline (Figure 5.6A). While control mice respond somewhat more at the cocaine lever than an inactive lever, reflecting partial reinforcement, this low dose of cocaine is self-administered at very high rates after initial acquisition in ΔFosB mice. When the initial training dose is increased to 250 μg/kg/injection, both ΔFosB mice and their controls acquire and maintain cocaine self-administration at similar rates (Figure 5.6B). The higher dose is self-administered at lower rates in ΔFosB mice since the injections produce prolonged effects as discussed earlier. Together, these results indicate that mice overexpressing ΔFosB in dynorphin-containing striatal neurons exhibit enhanced pharmacological sensitivity to cocaine reinforcement, although this effect alone has not been directly related to addicted phenotypes in animal models of cocaine addiction as discussed above.

Following the initial acquisition phase, animals receive further acquisition training at a higher injection dose (500 μg/kg/injection) for 5 days to engender acquisition and stabilization of self-administration in most mice for subsequent use in post-acquisition tests (Figure 5.4). Using this acquisition procedure, the latency to acquire cocaine self-administration can be determined based on the number of test sessions required to achieve >15 cocaine injections for three consecutive test sessions, each with a 3:1 ratio of active to inactive lever presses. These criteria ensure that animals acquire stable self-administration and lever discrimination. ΔFosB-expressing mice initially trained on the low 125 μg/kg/injection dose acquire cocaine self-administration with an average latency of ~5 test sessions, while their littermate controls average ~8 test sessions, indicating that the higher dose tested in days 6–10 is needed in controls (Figure 5.6A). In contrast, both study groups acquire cocaine self-administration with an average latency of about seven test sessions when initially trained on the 250 μg/kg/injection dose (Figure 5.6B). ΔFosB expressing mice stabilize their cocaine intake at control levels when self-administering the higher injection dose, indicating that increases in self-administration at the subthreshold dose are not related to a generalized rate-enhancing effect of ΔFosB.

In mice that acquire cocaine self-administration, the response requirement is gradually raised from FR1 to FR5 over 8–10 days, and until cocaine intake varies <15% of the mean of three consecutive test sessions (Figure 5.4). Once criteria for stabilization are met, mice are tested in a between-session dose-response procedure over 12 daily test sessions on the FR5 schedule, with each dose tested for two consecutive test sessions. The dose-response curve is generated using data from the second test day at each dose, which, along with the FR5 schedule, reduces the potential for extinction behavior to influence drug intake. As discussed above, addiction-related changes are indicated by a vertical shift in the inverted U-shaped self-administration dose-response curve on FR reinforcement schedules, reflecting an escalation in preferred levels of drug intake. As shown in Figure 5.7A, both

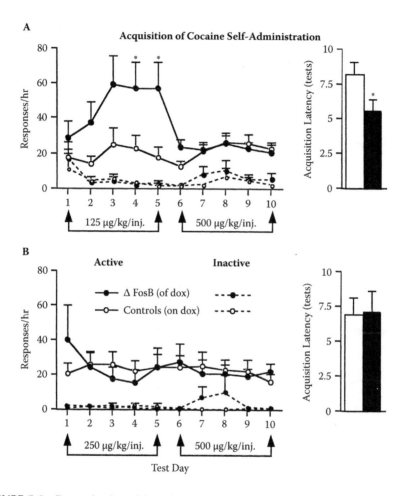

FIGURE 5.6 Determination of dose thresholds for acquisition of cocaine self-administration in mice with inducible striatal cell type–specific transgenic overexpression of ΔFosB. (A) Overexpression of ΔFosB facilitates acquisition of cocaine self-administration (FR1) at a low-threshold dose of cocaine (125 μg/kg/injection), (B) but not at a higher suprathreshold dose (250 μg/kg/injection), compared with littermate controls maintained on doxycycline. The number of responses at the active lever that delivers cocaine injections (solid lines) and inactive lever presses (dashed lines) is shown at left. The number of test sessions (latency) to achieve acquisition criteria is shown at right (>15 injections for 3 consecutive days, each with a 3:1 ratio of active:inactive lever presses). Each dose is tested for 5 days, followed by a higher training dose (500 μg/kg per injection) for days 6–10 to demonstrate a capacity for acquisition in all mice used in the analysis. Asterisks indicate that ΔFosB mice differ from littermate controls maintained on doxycycline for threshold dose cocaine self-administration ($P < 0.05$). (From Colby et al., *J. Neurosci.*, 23, 2488, 2003. With permission.)

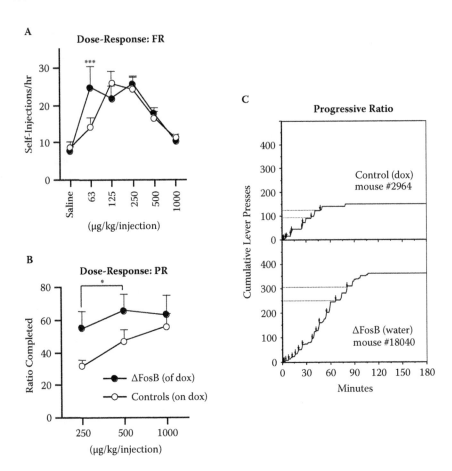

FIGURE 5.7 Determination of self-administration dose-response curves on fixed (FR) and progressive (PR) ratio reinforcement schedules in mice stabilized on FR5 (500 µg/kg/injection) prior to testing. (A) In FR dose-response testing, striatal cell–specific overexpression of ΔFosB increases sensitivity to the reinforcing effects of a low-threshold dose of cocaine but does not alter preferred levels of cocaine intake at higher doses on the descending limb of the curve that support self-administration in littermate controls maintained on doxycycline. Asterisks indicate that the 63 µg/kg/injection dose differs from saline. (B) In contrast, mice overexpressing ΔFosB exert greater effort to maintain self-administration on the PR schedule of reinforcement at suprathreshold doses, as reflected by a higher ratio of responses per injection achieved before mice quit self-administration altogether (ratio completed). (C) An example of this effect is shown in cumulative active lever-response records for representative mice showing that a ΔFosB mouse achieves a response-to-injection ratio of 56 lever presses before cessation of cocaine self-administration (vertical distance between dotted lines), whereas a control mouse achieves only 30 lever presses/injection before quitting self-administration (250 µg/kg per injection). The arrows denote times where injections are earned. Self-administration differences under FR and PR reinforcement schedules with similar cocaine doses indicate that cocaine intake and cocaine-seeking behavior are differentially regulated by ΔFosB. (From Colby et al., *J. Neurosci.*, 23, 2488, 2003. With permission.)

ΔFosB overexpressing mice and their littermate controls self-administer cocaine with typical inverted U-shaped dose-response curves. However, only ΔFosB mice maintain cocaine self-administration above saline self-administration rates at the lowest cocaine dose tested (63 µg/kg/injection), consistent with enhanced sensitivity to low-dose cocaine reinforcement shown in acquisition testing. The potential influence of ΔFosB on response perseveration is negated by the fact that saline substitution reduces responses to similar levels in both study groups, and by similar extinction latencies from food-reinforced behavior as discussed above. However, ΔFosB expression fails to cause a vertical shift in cocaine intake on the descending limb of the curve. Thus, ΔFosB apparently does not contribute to addiction-related escalation in preferred levels of cocaine intake.

Following FR dose-response testing in this procedure, mice return to the training dose (500 µg/kg/injection) until FR5 self-administration is stable for at least 3 days. Mice are then subjected to dose-response testing on the PR reinforcement schedule over 6 days, with each of three injection doses available for two consecutive test sessions (Figure 5.4). As discussed earlier, the PR schedule determines the maximum number of drug-seeking response an animal will perform to maintain drug self-administration, and addiction-related changes are indicated by an increase in the highest ratio of lever-press responses/injection achieved before cessation of self-administration behavior (Mendrek et al. 1998; Piazza et al. 2000; Suto et al. 2003). The degree of effort exerted by mice to maintain cocaine self-administration is thought to reflect the incentive strength of cocaine, or reinforcing efficacy. It is important to employ injection doses that are readily self-administered on FR reinforcement schedules; otherwise, differences in PR responding could simply reflect changes in dose threshold sensitivity rather than the motivational strength of cocaine.

Although the injection dose of cocaine is inversely related to response rates on FR schedules, reflecting the dose-related duration of cocaine effects, the same dose range produces dose-related increases in response rates on PR schedules (e.g., Figure 5.7B, controls), reflecting the dose-related efficacy of cocaine reinforcement. Thus, higher injection doses are more reinforcing since mice will exert greater effort to obtain them on PR schedules, but the longer duration of action reduces the overall rate of self-administration on FR schedules when responding is more tightly coupled to injection delivery. Despite the fact that ΔFosB and control mice self-administer similar amounts of cocaine at suprathreshold doses on FR reinforcement schedules, the ΔFosB mice exert far greater effort to self-administer the same cocaine doses on the PR schedule (Figure 5.7B). The enhanced motivation for cocaine in ΔFosB mice is illustrated in cumulative response records for representative mice depicted in Figure 5.7C. As the response requirement for cocaine injections progressively increases with each successive cocaine injection (arrows), the control mouse quits self-administration after achieving a response/injection ratio of only 30 lever presses (vertical distance between dotted lines). In contrast, the mouse overexpressing ΔFosB continues self-administering cocaine until a response/injection ratio of 56 lever presses is achieved. ΔFosB mice complete higher response/injection ratios at the 250 and 500 µg/kg/injection doses, but convergence at the highest injection dose probably reflects a performance ceiling. Moreover, similar maximal response/ injection ratios in both groups indicate that enhanced PR responding at lower doses is not due to a greater capacity for lever-press performance. Thus, overexpression of ΔFosB in dynorphin-containing striatal neurons increases cocaine-seeking behavior when reinforcement is withheld.

This serial testing procedure incorporates several measures of operant behavior with natural and intravenous drug rewards. A major limiting factor in conducting mouse self-administration studies is the longevity of intravenous catheter patency, and the viability of catheterized mice self-administering intravenous drugs. The genetic background can also influence both catheter patency and viability. In our studies, mouse catheters typically remain patent for 5–8 weeks, and mice that encounter catheter failures during testing can receive a second catheter in the contralateral jugular vein in a second minor surgical procedure. An important aspect of the serial testing procedure described above is that far more mice are tested in initial acquisition of self-administration than ultimately are needed for dose-response analysis on FR and PR reinforcement schedules. Thus, initial acquisition is tested with two threshold-spanning doses in separate study groups, and these additional mice help to offset viability and catheter failure issues that occur with greater frequency in postacquisition self-administration testing.

In more recent studies, we compared the effects of cyclic AMP activation in two distinct striatal cell types on cocaine self-administration on FR and PR reinforcement schedules. A Tet-Op regulated transgene for the Gsα subunit of stimulatory G proteins was overexpressed as a method to upregulate cyclic AMP signaling, a neuroadaptation common to chronic exposure to several drugs of abuse (Self and Nestler 1998). We selected two different tTA transgenic mouse lines to drive expression of the Gsα subunit primarily in either dynorphin- or enkephalin-containing striatal neurons that represent the direct or indirect striatal output pathways, respectively. The tTA mouse lines were chosen based on their previous history of restricting other transgenic expression to these cell types. While the selection of tTA driver lines does not ensure cell-specific expression patterns with every transgene, differential cell-specific expression was indicated by an increase in dynorphin mRNA (a cyclic AMP-regulated gene) only in the line targeting the dynorphin cell type, and by colabeling of the transgene in either dynorphin- or enkephalin-containing striatal neurons in each tTA driver line. These studies suggest that transgenic expression of the Gsα subunit produces very different effects on cocaine self-administration behavior depending on the striatal cell type expressing the transgene (Ruiz et al. 2005).

Transgenic expression of Gsα in dynorphin-containing neurons produces a tolerance-like escalation in cocaine self-administration and a vertical shift in the dose-response curve when access to cocaine is not restricted on FR schedules. However, Gsα expression in this cell type fails to alter cocaine-seeking behavior on the PR reinforcement schedule when successive cocaine injections require greater effort. Conversely, mice with Gsα expression in enkephalin-containing neurons exert far greater effort to obtain cocaine injections on the PR reinforcement schedule, but do not escalate their cocaine intake on the less demanding FR schedule. These results suggest that upregulation in cyclic AMP signaling with chronic drug use would differentially contribute to escalating cocaine intake and the propensity for cocaine-seeking behavior depending on the striatal cell type where cyclic AMP activity is elevated. In contrast to cell-specific transgenic expression, pharmacological approaches that produce generalized activation of cyclic AMP signaling in both cell types led to increased cocaine self-administration on both FR and PR reinforcement schedules (Self et al. 1998; Lynch and Taylor 2005).

5.5 CELL-SPECIFIC GENETIC DELETION IN MOUSE SELF-ADMINISTRATION STUDIES

Genetic deletion, or knockout, is an extremely useful loss-of-function approach to study the necessary role of neuronal proteins in behavior. In some cases, genetic deletion studies have led to provocative and unexpected results. For example, early work with dopamine transporter (DAT) knockout mice challenged the theory that DAT is a major target for the reinforcing effects of cocaine, despite numerous and seemingly conclusive evidence that DAT involvement is critical (e.g., Kuhar et al. 1991). Interestingly, DAT knockout mice will learn to self-administer cocaine, although they acquire self-administration more slowly than wild-type controls (Rocha et al. 1998a). Initial studies showed that cocaine fails to elevate extracellular dopamine levels in the dorsal neostriatum of DAT knockouts (Rocha et al. 1998a), suggesting that DAT is not a major pharmacological target for cocaine reinforcement. However, subsequent studies found that cocaine does increase dopamine levels in the shell subregion of the nucleus accumbens in DAT knockout mice (Carboni et al. 2001; Mateo et al. 2004), but through circuitous routes that are only revealed in the absence of DAT. One mechanism may involve cocaine's ability to block dopamine uptake by the norepinephrine transporter that is prominently expressed in nucleus accumbens shell, but not the neostriatum (Berridge et al. 1997). Another mechanism may involve cocaine-induced increases in norepinephrine in the prefrontal cortex that indirectly led to dopamine increases in the nucleus accumbens (Goeders and Smith 1993), most likely through activation of excitatory cortical afferents to the ventral tegmental area. Finally, cocaine blockade of serotonin transporters and activation of serotonin receptors in the ventral tegmental area may stimulate dopamine release in the nucleus accumbens of DAT knockout mice (Mateo et al. 2004). Collectively, these data illustrate secondary but potentially important mechanisms of cocaine-induced dopamine elevations that are unmasked by DAT knockout. However, results from DAT knockout mice do not exclude DAT as a major target for cocaine reinforcement.

Most embryonic knockouts are susceptible to developmental and homeostatic compensation, and the anatomically diffuse loss of brainwide protein function can indirectly alter neural network activity-regulating behavior in a manner that complicates interpretation. For example, embryonic DAT knockouts show decreased expression of D1 and D2 receptors, a homeostatic compensation to elevated dopamine levels, but also a reorganization involving an increase in enkephalin-containing nucleus accumbens neurons that express D3 receptors that normally are expressed primarily in dynorphin-containing neurons (Fauchey et al. 2000). DAT knockout mice also show a loss of autoreceptor function in midbrain dopamine neurons (Jones et al. 1999).

Inducible knockout technology allows investigators to bypass potential developmental effects, and combines the anatomical specificity of transgenic expression systems. Inducible knockout is accomplished by flanking (floxing) targeted genes with the insertion of 34 base-pair loxP sequences, combined with inducible transgenic expression of the enzyme Cre recombinase to excise genetic material between the loxP sites in mature adult neurons (for review, see Branda and Dymecki 2004). The region and cell type where genes are deleted is determined by the anatomical expression pattern of transgenic Cre recombinase. This approach avoids lethality often associated with developmental perturbations or widespread loss of protein function, and can utilize both targeted and randomly generated transgenic approaches to obtain the desired anatomical profile of gene deletion as described above.

An alternative approach uses viral-mediated gene transfer to locally express Cre recombinase when the viral vector is infused into a specific brain site of mice possessing a floxed gene (Figure 5.8). This approach involves the powerful combination of a highly localized knockout with temporal control over genetic deletion in adult developmental stages. Certain modifications in viral vectors and the selection of appropriate viral strains conveys selective expression in neurons with little or no potential for retrograde infection of afferents inputs to targeted brain regions, thereby limiting genetic deletion to local neuronal cell bodies (Chamberlin et al. 1998; Davidson and Breakefield 2003). The viral-mediated knockout approach can be used to either delete or activate protein function, depending on whether the targeted gene itself or an inhibitory promoter region is floxed (Brooks et al. 1997; South et al. 2003). Prominent behavioral effects with viral-mediated Cre recombinase

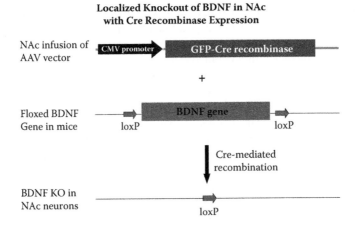

FIGURE 5.8 Region-specific knockout of BDNF in adult mice. The powerful combination of viral vector and inducible knockout technology circumvents problems relating to developmental compensation and diffuse transgenic expression patterns. Using intracerebral infusions of an AAV viral vector that expresses Cre recombinase in mice with a floxed BDNF gene, highly localized BDNF knockout is produced. The BDNF gene is flanked by the insertion of loxP sites through recombinant technology in embryonic stem cells that are injected into a mouse embryo before implantation in a pseudopregnant female. The loxP sites are excised by recombinase expressed after the viral vector is infused in the adult mouse.

expression are likely because even low expression levels for relatively brief periods of time can effectively and irreversibly delete floxed sequences in infected neurons.

We utilized a similar approach to study the role of brain-derived neurotrophic factor (BDNF) derived exclusively from nucleus accumbens neurons on cocaine self-administration. BDNF is induced and released in the nucleus accumbens with daily cocaine use (Graham et al. 2007), but the source of BDNF protein could involve afferent input originating in the prefrontal cortex or ventral tegmental area (Altar et al. 1997; Guillin et al. 2001). An adeno-associated viral vector (AAV) expressing Cre recombinase with green fluorescent protein (GFP), or GFP alone as a control, was infused into the nucleus accumbens of floxed BDNF mice. As depicted in Figure 5.9A, BDNF mRNA is virtually undetectable by quantitative real-time polymerase chain reaction (RT-PCR) in laser-captured neurons expressing Cre recombinase when compared to neurons infected with GFP alone. Localized BDNF knockout also substantially reduces BDNF protein levels by 46% in the nucleus accumbens (Figure 5.9B), suggesting that a substantial proportion of BDNF protein originates in nucleus accumbens neurons in addition to afferent sources. Importantly, the AAV strain (AAV2) does not express by retrograde infection of neuronal afferents (Chamberlin et al. 1998), as confirmed by a lack of GFP expression in distal brain regions that project to the nucleus accumbens.

We studied the effects of inducible BDNF knockout in nucleus accumbens neurons on operant learning capacity and cocaine self-administration in floxed BDNF mice receiving nucleus accumbens infusions of AAV-Cre recombinase. In untrained mice maintained on a restricted diet, localized BDNF knockout in nucleus accumbens

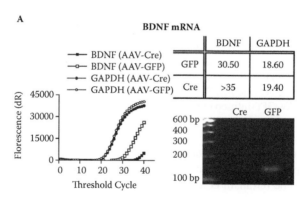

FIGURE 5.9 Inducible localized knockout of BDNF in nucleus accumbens (NAc) neurons reduces cocaine reinforcement. (A) Verification of BDNF knockout by AAV-Cre infection and recombination in mice with lox P sites flanking the BDNF coding region (exon 5). Quantitative RT-PCR in laser-captured GFP-positive NAc neurons shows the elimination of BDNF mRNA in cells infected with AAV-Cre but not in cells infected with AAV-GFP alone (control). Representative amplification curves show no differences in GAPDH mRNA as an internal control (mean of three replicates). Mean cycle threshold values for determination of BDNF and GAPDH mRNA levels are shown (table) and represent pooled cells from three to four animals/treatment. BDNF RT-PCR products (150 bp) separated on agarose gel are readily detectable in cells infected with AAV-GFP, but not in cells infected with AAV-Cre. From Graham et al., *Nat Neurosci*, 10, 1029, 2007. With permission.

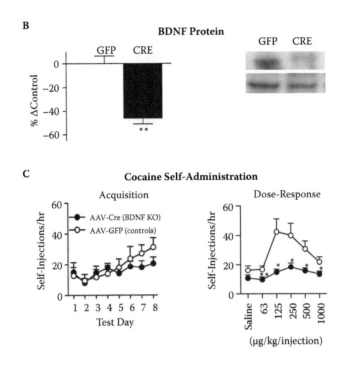

FIGURE 5.9 (continued). (B) Immunoblots from tissue infected with AAV-Cre show a 46% reduction in BDNF protein levels compared with AAV-GFP-infected tissue. Data are expressed as mean ± sem. **$P < 0.01$, compared with AAV-GFP-treated animals. (C) Localized knockout of BDNF in NAc neurons has no effect on initial acquisition of cocaine self-administration on a less demanding FR1 schedule (left panel). Following acquisition and stabilization on FR5, the dose-response curve for cocaine self-administration is shifted downward by BDNF knockout in NAc neurons, indicating a reduction in cocaine reinforcement under the more demanding FR5 schedule. Asterisks indicate *$P < 0.05$ and **$P < 0.01$ compared with AAV-GFP-infused mice. (From Graham et al., *Nat. Neurosci.*, 10, 1029, 2007. With permission.)

neurons has no effect on spontaneous lever-press behavior and no effect on the acquisition of lever-press behavior reinforced by 25-mg sucrose pellets on an FR3 reinforcement schedule (Graham et al. 2007). These data indicate that mechanisms of operant learning remain intact with localized BDNF knockout in nucleus accumbens neurons. Following lever-press training, these mice were allowed to self-administer intravenous cocaine injections at a suprathreshold training dose on an FR1 reinforcement schedule. Figure 5.9C shows that both localized BDNF knockout mice and control mice learn to self-administer cocaine with similar acquisition latencies of ~6 days to meet criteria (number of sessions until >15 self-injections for 3 consecutive days with ≥ 3:1 active:inactive lever-response ratio). This finding indicates that localized loss of BDNF expression does not completely attenuate cocaine reinforcement. However, when the response requirement is raised to a more demanding FR5 schedule, the inverted U-shaped dose-response curve is generally flat and shifted downward in BDNF knockout mice when compared with GFP-expressing controls (Figure 5.9C). These data indicate that BDNF originating in nucleus accumbens neurons enhances cocaine reinforcement, an effect revealed under the more demanding FR5 reinforcement schedule. Moreover, BDNF derived from nucleus accumbens neurons evidently is critical for addiction-like vertical shifts in the FR self-administration dose-response curve. Such dramatic and behaviorally selective changes in cocaine self-administration with anatomically restricted BDNF deletion provides strong evidence that dynamic increases in nucleus accumbens-derived BDNF with cocaine use play a necessary role in the transition to more addicted biological states.

5.6   CONCLUDING REMARKS

Previously, self-administration studies have been constrained by the availability of selective pharmacological ligands that target a limited array of neurotransmitter receptors, or proteins with enzymatic activity. Although information gained from this approach has clearly advanced our understanding of the pharmacological mechanisms regulating drug intake and drug reinforcement, many neuroadaptations to chronic drug use involve intracellular proteins inaccessible to such pharmacological manipulation. Pharmacological ligands typically are given as a pretreatment immediately prior to self-administration tests, which is entirely appropriate for delineating the role of neurotransmitter receptors in regulating self-administration behavior. However, such brief and acute modulation of receptor targets fails to mimic the enduring nature of neurobiological changes to chronic drug use. Thus, neuroadaptations in specific proteins may lead to secondary homeostatic compensation or dysregulation in downstream cellular processes. Although these secondary compensatory changes can interfere with determining the role of specific proteins in the acute regulation of drug self-administration behavior, they encompass a broader spectrum of changes that would accompany enduring up- or downregulation in these proteins with chronic drug use. Thus, modern transgenic and gene deletion approaches are better suited to experimentally reproduce the enduring brain changes that occur with the development of addicted biological states.

The application of traditional molecular genetic approaches to analysis of self-administration behavior also is hampered by a lack of temporal and anatomical control over target protein expression. However, in some cases, traditional knockout mice have been useful to gain initial information on receptors that lack available ligands, such as a role for M5 muscarinic cholinergic receptors in cocaine self-administration (Thomsen et al. 2005). Discoveries made with knockout mice also have generated interest in novel targets for potential therapeutic intervention. For example, the failure of mGluR5 glutamate receptor knockout mice to self-administer cocaine has generated interest in mGluR5 glutamate receptor antagonists as potential treatments for cocaine abuse (Chiamulera et al. 2001; Kenny et al. 2003). In other cases, dopamine D2 receptor knockout mice have been used to confirm the selectivity of D2 and D3 receptor ligands and the role of the D2 receptor subtype in cocaine self-administration behavior (Caine et al. 2002). Finally, studies with DAT knockout mice clearly established novel mechanisms of cocaine action and potential secondary mechanisms of cocaine reinforcement (Rocha et al.1998a).

In contrast, the role of specific molecular adaptations to chronic drug use in the etiology of addiction requires better anatomical control over regional and cell-specific expression patterns to clearly establish definitive cause–effect relationships with addictive behavior. The application of modern genetic technology with better temporal and anatomical control over protein expression represents an enormous advance for drug self-administration studies aimed at determining significant functional interactions between brain and behavior. Furthermore, inducible and cell-specific genetic manipulation already has been used to discover important differences in the neurobiological mechanisms that regulate drug-taking and drug-seeking behaviors when drugs are self-administered on either FR or PR reinforcement schedules. These differences illustrate the complexity of motivational systems that regulate drug self-administration behavior, and they underscore the utility of self-administration procedures to fully appreciate this complexity.

REFERENCES

Ahmed, S. H., and G. F. Koob. (1998). Transition from moderate to excessive drug intake: change in hedonic set point. *Science* 282: 298–300.

Altar, C. A., Cai, N., Bliven, T., Juhasz, M., Conner, J. M., Acheson, A. L., Lindsay, R. M., and S. J. Wiegand. (1997). Anterograde transport of brain-derived neurotrophic factor and its role in the brain. *Nature* 389:856–860.

American Psychiatric Association. (2000). *Diagnostic and Statistical Manual of Mental Disorders, 4th ed., Text Revision.* Washington, D.C.: American Psychiatric Association.

Berridge, K. C., and T. E. Robinson. (1998). What is the role of dopamine in reward: hedonic impact, reward learning, or incentive salience? *Brain Res Rev* 28:309–369.

Berridge, C. W., Stratford, T. L., Foote, S. L., and A. E. Kelley. (1997). Distribution of dopamine beta-hydroxylase-like immunoreactive fibers within the shell subregion of the nucleus accumbens. *Synapse* 27:230–241.

Branda, C. S., and S. M. Dymecki. (2004). Talking about a revolution: the impact of site-specific recombinases on genetic analyses in mice. *Dev Cell* 6:7–28.

Brooks, A. I., Muhkerjee, B. Panahian, N., Cory-Slechta, D., and H. J. Federoff. (1997). Nerve growth factor somatic mosaicism produced by herpes virus-directed expression of cre recombinase. *Nat Biotechnol* 15:57–62.

Caine, S. B., Negus, S. S., Mello, N. K., Patel, S., Bristow, L., Kulagowski, J., Vallone, D., Saiardi, A., and E. Borrelli. (2002). Role of dopamine D2-like receptors in cocaine self-administration: studies with D2 receptor mutant mice and novel D2 receptor antagonists. *J Neurosci* 22:2977–2988.

Carboni, E., Spielewoy, C., Vacca, C., Nosten-Bertrand, M., Giros, B., and G. Di Chiara. (2001). Cocaine and amphetamine increase extracellular dopamine in the nucleus accumbens of mice lacking the dopamine transporter gene. *J Neurosci* 21:RC141:1–4.

Carney, J. M., Landrum, R. W., Cheng, M. S., and T. W. Seale. (1991). Establishment of chronic intravenous drug self-administration in the C57BL/6J mouse. *Neuroreport* 2:477–480.

Carter, D. A. (2004). Comprehensive strategies to study neuronal function in transgenic animal models. *Biol Psychiatry* 55:785–788.

Chamberlin, N. L., Du, B., de Lacalle, S., and C. B. Saper. (1998). Recombinant adeno-associated virus vector: use for transgene expression and anterograde tract tracing in the CNS. *Brain Res* 793:169–175.

Chen, J.-S., Kelz, M. B., Zeng, G., Sakai, N., Steffan, C., Shockett, P. E., Picciotto, M. R., Duman, R. S., and E. J. Nestler. (1998) Transgenic animals with inducible, targeted gene expression in brain. *Mol Pharm* 54:495–503.

Chiamulera, C., Epping-Jordan, M. P., Zocchi, A., Marcon, C., Cottiny, C., Tacconi, S., Corsi, M., Orzi, F., and F. Conquet. (2001). Reinforcing and locomotor stimulant effects of cocaine are absent in mGluR5 null mutant mice. *Nat Neurosci* 4:873–874.

Colby, C. R., Whisler, K., Steffen, C., Nestler, E. J., and D. W. Self. (2003). Striatal cell type-specific overexpression of ΔFosb enhances incentive for cocaine. *J Neurosci* 23:2488–2493.

Collins, R. J., Weeks, J. R., Cooper, M. M., Good, P. I., and R. R. Russell. (1984). Prediction of abuse liability of drugs using iv self-administration by rats. *Psychopharmacology* 82:6–13.

David, V., Gold, L. H., Koob, G. F., and P. Cazala. (2001a). Anxiogenic-like effects limit rewarding effects of cocaine in Balb/Cbyj Mice. *Neuropsychopharmacology* 24:300–318.

David, V., Polis, I., McDonald, J., and L. H. Gold. (2001b). Intravenous self-administration of heroin/cocaine combinations (speedball) using nose-poke or lever-press operant responding in mice. *Behav Pharmacol* 12:25–34.

Davidson, B. L., and X. O. Breakefield. (2003) Viral vectors for gene delivery to the nervous system. *Nat Rev Neurosci* 4:353–364.

Deneau, G., Yanagita, T., and M. H. Seevers. (1969). Self-administration of psychoactive substances by the monkey. *Psychopharmacologia* 16:30–48.

Deroche, V., Caine, S. B., Heyser, C. J., Polis, I., Koob, G. F., and L. H. Gold. (1997). Differences in the liability to self-administer intravenous cocaine between C57bl/6 x Sjl and Balb/Cbyj mice. *Pharmacol Biochem Behav* 57:429–440.

Deroche, V., Le Moal, M., and P. V. Piazza. (1999). Cocaine self-administration increases the incentive motivational properties of the drug in rats. *Eur J Neurosci* 11:2731–2736.

Deroche-Gamonet, V., Belin, D., and P. V. Piazza. (2004). Evidence for addiction-like behavior in the rat. *Science* 305:1014–1017.

Edwards, S., Whisler, K. N., Fuller, D. C., Orsulak, P. J., and D. W. Self. (2007). Addiction-related alterations in D_1 and D_2 dopamine receptor behavioral responses following chronic cocaine self-administration. *Neuropsychopharmacology* 32:354–366.

Fauchey, V., Jaber, M., Caron, M. G., Bloch, B., and C. Le Moine. (2000). Differential regulation of the dopamine D1, D2 and D3 receptor gene expression and changes in the phenotype of the striatal neurons in mice lacking the dopamine transporter. *Eur J Neurosci* 12:19–26.

Fuchs, R. A., See, R. E., and L. D. Middaugh. (2003). Conditioned stimulus-induced reinstatement of extinguished cocaine seeking in C57bl/6 mice: a mouse model of drug relapse. *Brain Res* 973:99–106.

Goeders, N. E., and J. E. Smith. (1993). Intracranial cocaine self-administration into the medial prefrontal cortex increases dopamine turnover in the nucleus accumbens. *J Pharmacol Exp Ther* 265:592–600.

Gong, S., Zheng, C., Doughty, M. L., Losos, K., Didkovsky, N., Schambra, U. B., Nowak, N. J., Joyner, A., Leblanc, G., Hatten, M. E., and N. Heintz. (2003). A gene expression atlas of the central nervous system based on bacterial artificial chromosomes. *Nature* 425:917–925.

Graham, D. L., Edwards, S., Bachtell, R. K., DiLeone, R. J., Rios, M., and D. W. Self. (2007). Dynamic BDNF activity in nucleus accumbens with cocaine use increases self-administration and relapse. *Nat Neurosci* 10:1029–1037.

Grahame, N. J., and C. L. Cunningham. (1995). Genetic differences in intravenous cocaine self-administration between C57BL/6J and DBA/2J mice. *Psychopharmacology (Berl)* 122:281–291.

Grahame, N. J., Phillips, T. J., Burkhart-Kasch, S., and C. L. Cunningham. (1995). Intravenous cocaine self-administration in the C57BL/6J mouse. *Pharmacol Biochem Behav* 51:827–834.

Guillin, O., Diaz, J., Carroll, P., Griffon, N., Schwartz, J. C., and P. Sokoloff. (2001). BDNF controls dopamine D3 receptor expression and triggers behavioural sensitization. *Nature* 411:86–89.

Highfield, D. A., Mead, A. N., Grimm, J. W., Rocha, B. A., and Y. Shaham. (2002). Reinstatement of cocaine seeking in 129X1/SvJ mice: effects of cocaine priming, cocaine cues, and food deprivation. *Psychopharmacology* 161:417–424.

Jaffe, J. H., Cascella, N. G., Kumor, K. M., and M. A. Sherer. (1989). Cocaine-induced cocaine craving. *Psychopharmacology* 97:59–64.

Johanson, C. E., and M. W. Fischman. (1989). The pharmacology of cocaine related to its abuse. *Pharmacol Rev* 41:3–52.

Jones, S. R., Gainetdinov, R. R., Hu, X. T., Cooper, D. C., Wightman, R. M., White, F.J., and M. G. Caron. (1999). Loss of autoreceptor functions in mice lacking the dopamine transporter. *Nat Neurosci* 2:649–655.

Kelz, M. B., Chen, J., Carlezon, Jr., W. A., Whisler, K., Gilden, L., Beckmann, A. M., Steffen, C., Zhang, Y. J., Marotti, L., Self, D. W., Tkatch, T., Baranauskas, G., Surmeier, D. J., Neve, R. L., Duman, R. S., Picciotto, M. R., and E. J. Nestler. (1999). Expression of the transcription factor ΔFosB in the brain controls sensitivity to cocaine. *Nature* 401:272–276.

Kenny, P. J., Paterson, N. E., Boutrel, B., Semenova, S., Harrison, A. A., Gasparini, F., Koob, G. F., Skoubis, P. D., and A. Markou. (2003). Metabotropic glutamate 5 receptor antagonist MPEP decreased nicotine and cocaine self-administration but not nicotine and cocaine-induced facilitation of brain reward function in rats. *Ann N Y Acad Sci* 1003:415–418.

Kruzich, P. J. (2007). Does response-contingent access to cocaine reinstate previously extinguished cocaine-seeking behavior in C57BL/6J mice? *Brain Res* 1149:165–171.

Kuhar, M. J., Ritz, M. C., and J. W. Boja. (1991). The dopamine hypothesis of the reinforcing properties of cocaine. *Trends Neurosci* 14: 299–302.

Kuzmin, A., and B. Johansson. (2000). Reinforcing and neurochemical effects of cocaine: differences among C57, Dba, and 129 mice. *Pharmacol Biochem Behav* 65:399–406.

Lynch, W. J., and J. R. Taylor. (2005). Persistent changes in motivation to self-administer cocaine following modulation of cyclic AMP-dependent protein kinase A (PKA) activity in the nucleus accumbens. *Eur J Neurosci* 22:1214–1220.

Mateo, Y., Budygin, E. A., John, C. E., and S. R. Jones. (2004). Role of serotonin in cocaine effects in mice with reduced dopamine transporter function. *Proc Natl Acad Sci U S A* 101:372–377.

Mendrek, A., Blaha, C. D., and A. G. Phillips. (1998). Pre-exposure of rats to amphetamine sensitizes self-administration of this drug under a progressive ratio schedule. *Psychopharmacology* 135:416–422.

Moratalla, R., Elibol, B., Vallejo, M. and A. M. Graybiel. (1996) Network-level changes in expression of inducible Fos-Jun proteins in the striatum during chronic cocaine treatment and withdrawal. *Neuron* 17:147–156.

Morozov, A., Kellendonk, C., Simpson, E., and F. Tronche. (2003). Using conditional mutagenesis to study the brain. *Biol Psychiat* 54:1125–1133.

Morse, A. C., Erwin, V. G., and B. C. Jones. (1993). Strain and housing affect cocaine self-selection and open-field locomotor activity in mice. *Pharmacol Biochem Behav* 45:905–912.

Nye, H. E., Hope, B. T., Kelz, M. B., Iadarola, M. and E. J. Nestler. (1995) Pharmacological studies of the regulation of chronic FOS-related antigen induction by cocaine in the striatum and nucleus accumbens. *J Pharmacol Exp Ther* 275:1671–1680.

Palmer, C. A., Lubon, H., and J. L. McManaman. (2003). Transgenic mice expressing recombinant human protein C exhibit defects in lactation and impaired mammary gland development. *Transgenic Res* 12:283–292.

Piazza, P. V., Deroche-Gamonent, V., Rouge-Pont, F., and M. Le Moal. (2000). Vertical shifts in self-administration dose-response functions predict a drug-vulnerable phenotype predisposed to addiction. *J Neurosci* 20:4226–4232.

Robbins, S. J., Ehrman, R. N., Childress, A. R., and C. P. O'Brien. (1997). Relationships among physiological and self-report responses produced by cocaine-related cues. *Addict Behav* 22:157–167.

Robinson, T. E., and K. C. Berridge. (1993). The neural basis of drug craving: an incentive-sensitization theory of addiction. *Brain Res Rev* 18:247–291.

Rocha, B. A., Fumagalli, F., Gainetdinov, R .R., Jones, S. R., Ator, R., Giros, B., Miller, G. W., and M. G. Caron. (1998a). Cocaine self-administration in dopamine-transporter knockout mice. *Nat Neurosci* 1:132–137.

Rocha, B. A., Odom, L. A., Barron, B. A., Ator, R., Wild, S. A., and M. J. Forster. (1998b). Differential responsiveness to cocaine in C57BL/6J and Dba/2j mice. *Psychopharmacology* 138:82–88.

Ruiz, E., C. Steffen, C., Abel, T., Nestler, E. J., and D. W. Self. (2005). Differential modulation of cocaine self-administration by expression of Gs-alpha subunits in putative direct and indirect striatal output neurons. *Soc Neurosci Abstr* 31:562.13.

Ruiz-Durantez, E., Hall, S. K., Steffen, C., and D. W. Self. (2006). Enhanced acquisition of cocaine self-administration by increasing percentages of C57/Bl/6j genes in mice with a nonpreferring outbred background. *Psychopharmacology* 186: 553–560.

Schuster, C. R., and T. Thompson. (1969). Self administration of and behavioral dependence on drugs. *Annu Rev Pharmacol* 9:483–502.

Sedivy, J., and A. Joyner. (1992). *Gene targeting*. New York: Freeman.

Self, D. W. (2005). Molecular and genetic approaches for behavioral analysis of protein function. *Biol Psychiat* 57:1479–1484.

Self, D. W., Genova, L. M., Hope, B.T., Barnhart, W. J., Spencer, J. J., and E. J. Nestler. (1998). Involvement of cAMP-dependent protein kinase in the nucleus accumbens in cocaine self-administration and relapse of cocaine-seeking behavior. *J Neurosci* 18:1848–1859.

Self, D. W., and E. J. Nestler. (1998). Relapse to drug seeking: neural and molecular mechanisms. *Drug Alcohol Depend* 51:49–60.

Sinha, R., Catapano, D., and S. O'Malley. (1999). Stress-induced craving and stress response in cocaine dependent individuals. *Psychopharmacology* 142:343–351.

South, S. M., Kohno, T., Kaspar, B. K., Hegarty, D., Vissel, B., Drake, C. T., Ohata, M., Jenab, S., Sailer, A. W., Malkmus, S., Masuyama, T., Horner, P., Bogulavsky, J., Gage, F. H., Yaksh, T. L., Woolf, C. J., Heinemann, S. F., and C. E. Inturrisi. (2003). A conditional deletion of the Nr1 subunit of the NMDA receptor in adult spinal cord dorsal horn reduces NMDA currents and injury-induced pain. *J Neurosci* 23:5031–5040.

Spanagel, R., and C. Sanchis-Segura. (2003). The use of transgenic mice to study addictive behavior. *Clin Neurosci Res* 3:325–331.

Stewart, J., de Wit, H., and R. Eikelboom. (1984). Role of unconditioned and conditioned drug effects in the self-administration of opiates and stimulants. *Psychol Rev* 91:251–268.

Suto, N., Tanabe, L. M., Austin, J. D., Creekmore, E., and P. Vezina. (2003). Previous exposure to VTA amphetamine enhances cocaine self-administration under a progressive ratio schedule in an NMDA, AMPA/Kainate, and metabotropic glutamate receptor-dependent manner. *Neuropsychopharmacology* 28:629–639.

Sutton, M. A., Karanian, D. A., and D. W. Self. (2000). Factors that determine a propensity for cocaine-seeking behavior during abstinence in rats. *Neuropsychopharmacology* 22:626–641.

Thomsen, M., and S. B. Caine. (2006). Cocaine self-administration under fixed and progressive ratio schedules of reinforcement: comparison of C57BL/6J, 129x1/Svj, and 129s6/Svevtac inbred mice. *Psychopharmacology* 184:145–154.

Thomsen, M., Woldbye, D. P., Wortwein, G., Fink-Jensen, A., Wess, J., and S. B. Caine. (2005). Reduced cocaine self-administration in muscarinic M5 acetylcholine receptor-deficient mice. *J Neurosci* 25:8141–8149.

Wells, T., and D. A. Carter. (2001). Genetic engineering of neural function in transgenic rodents: towards a comprehensive strategy? *J Neurosci Meth* 108:111–130.

Yan, Y., Nitta, A., Mizoguchi, H., Yamada, K., and T. Nabeshima. (2006). Relapse of methamphetamine-seeking behavior in C57BL/6J mice demonstrated by a reinstatement procedure involving intravenous self-administration. *Behav Brain Res* 168:137–143.

Wise, R. A. (2004). Drive, incentive, and reinforcement: the antecedents and consequences of motivation. *Nebraska Symp Motiv* 50:159–195.

6 Neuroeconomics
Implications for Understanding the Neurobiology of Addiction

Michael L. Platt, Karli K. Watson, Benjamin Y. Hayden, Stephen V. Shepherd, and Jeffrey T. Klein

CONTENTS

6.1 INTRODUCTION

Although still in the early stages, neuroeconomics—the union of ethology, economics, and neuroscience—offers a potentially powerful way to study the neural mechanisms underlying reward, punishment, and decision making, as well as the dysfunction of these systems in pathological conditions such as addiction. These neural processes interact in important ways with systems evaluating social context and uncertainty, and their study may lead to potent insights and testable predictions relevant to the neurobiology of addiction. Therefore, the neuroeconomic approach itself serves as a powerful new conceptual method that is likely to be critical for progress in understanding addictive behavior.

In the following review, we outline the neuroeconomic approach and illustrate how this conceptual method can inform studies of reward processing in general and addiction in particular. We will therefore discuss the neural processes that mediate motivation, those that translate endogenous rewards and punishments into behavioral decisions, and finally, the role of chemical perturbations of these processes in the etiology of addiction. We contend that the neurobiological basis of decision making in the context of addiction is critically interrelated with the neurobiology of economic decisions made by people and animals in their natural physical and social environments, and that both phenomena may be elucidated through a coherent suite of behavioral, neurophysiological, and pharmacological techniques.

We focus on the modulation of decision making by social context and economic risk. Although these two variables are not traditionally probed in neurobiological studies of addiction, they are likely to have important consequences for behavior. By necessity, such studies require the utilization of new behavioral techniques in which factors like social stimulation and economic risk are systematically varied, as well as analytical methods for relating these variables to neural function. We will therefore identify and describe examples of such techniques. Generally, we advocate the use of economic methods, which rely on the construction of preference functions and demand curves, to quantify and formally model context-dependent rewards and punishments in relation to the neurobiology of decision making and its pathology in addiction.

6.2 OVERVIEW OF THE NEUROBIOLOGY OF REWARD AND DECISION MAKING

6.2.1 Neurobiology of Reward and Motivation

Theoretical and empirical studies support the idea that, to the limits of physiological and cognitive constraints, behavioral choices optimize evolutionary fitness and are in this sense governed by economic principles. Accordingly, brains have become exquisitely specialized to attend key features of the environment, extract the fitness predictions of these features, and use this information to compute the optimal behavioral strategy. Following this logic, rewards can be considered proximate goals that, when acquired, tend to enhance survival and reproductive success. Similarly, avoiding punishment is a proximate goal that ultimately serves to enhance the long-term likelihood of survival and reproduction. These definitions extend the traditional psychological and neurobiological notions of reward and punishment, which typically are defined by elicited approach or avoidance, respectively.

Recent studies have revealed elementary properties of the neural systems that process rewards and punishments (Schultz and Dickinson 2000). Specifically, the neural circuits connecting midbrain dopamine neurons to the ventral striatum and prefrontal cortex appear to be crucial for processing information about rewards (Figure 6.1). For example, animals will work to receive current delivered via electrodes implanted in the dopaminergic ventral tegmental area (VTA) or in the medial forebrain bundle, which connects the VTA to the ventral striatum (Olds and Milner 1954). In fact, animals work for such intracranial self-stimulation to the exclusion

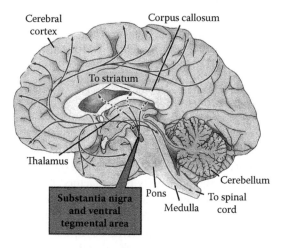

FIGURE 6.1 (Please see color insert following page 112.) The distribution of dopaminergic neurons and their projections in the human brain. The cell bodies of dopamine-containing neurons exist in clusters of cells (nuclei) located in the midbrain, and their axons project to widespread regions across the brain. Dopaminergic projection to cerebral cortex is heaviest in the frontal lobe, whereas projection to the occipital cortex is sparse. Subcortical structures such as the striatum and thalamus are also prominent dopaminergic targets. Adapted from Purves, D., Augustine. H. J., Fitzpatrick, D., Hall, W. C., LaMantia, A., McNamara, J. O., and Williams, S. M. 2004 *Neuroscience, Third Edition*. Sunderland, MA: Sinauer Associates. With permission.

of naturally rewarding fitness-enhancing behaviors like acquiring food or mating opportunities (White and Milner 1992).

Electrophysiological recordings from dopaminergic neurons show that these cells respond most strongly to primary rewards, such as food and water, when they are unexpected (Schultz 2000). When primary rewards are consistently predicted by other stimuli, as occurs during operant conditioning, dopaminergic responses shift from the predicted reward to the predictive stimulus (Schultz 2007). The size of the dopaminergic response scales both with the probability that a reward will follow a stimulus and with the magnitude of that predicted reward. Current evidence suggests that phasic bursts by dopamine neurons thus encode instantaneous reward predictions, producing a continuously updating fitness estimate that is formally equivalent to the "reward prediction error" term in computational models of learning (Rescorla and Wagner 1972; Schultz et al. 1997; Schultz and Dickinson 2000). According to this view, phasic dopamine (DA) signals are used both to learn about stimuli in the environment and to select profitable courses of action.

In addition to the ventral striatum, the medial prefrontal cortex (mPFC) and orbitofrontal cortex (OFC) appear to contribute to reward processing. Human functional imaging studies indicate that the OFC is activated by olfactory (Gottfried et al. 2002), gustatory (Kringelbach et al. 2003; O'Doherty et al. 2001), and tactile rewards (Rolls et al. 2003), as well as socially reinforcing stimuli such as beautiful or smiling faces (Aharon et al. 2001; O'Doherty et al. 2003). Lesions in this area in both humans (Hornak et al. 2003) and monkeys (Butter and Snyder 1972; Myers et al. 1973) result in decision-making impairment and inability to function appropriately within social groups.

Electrophysiological studies in both rats and monkeys additionally show that neurons in the OFC fire in response to and during anticipation of positive and negative outcomes (Hosokawa et al. 2007; Tremblay and Schultz 1999). Tremblay and Schultz (1999) demonstrated that OFC activity tracks the desirability of a reward relative to alternatives, and work by Critchley and Rolls (1996) revealed that some OFC neurons alter their responsiveness to gustatory rewards as satiety is reached. More formally, Padoa-Schioppa and Assad (2006) have demonstrated that activity of OFC neurons corresponds to monkeys' subjective preferences among desirable fluid rewards. These observations suggest that OFC neurons link predicted rewards to the current behavioral context, including both internal (i.e., motivation and satiety) and external (i.e., alternatives, opportunity costs) factors. In particular, it has been argued that the OFC transforms reward and punishment information into a common currency of subjective value along which any arbitrary options can be compared (Kringelbach et al. 2003; Montague and Berns 2002; Padoa-Schioppa and Assad 2006). Such computations are essential for effective decision making to take place. Together these observations implicate OFC in relating externally occurring events to internal states in a way that promotes survival and reproduction.

Several other brain areas are known to play critical roles in reward processing. The amygdala, in particular, has received extensive attention in recent years, especially for its role in mediating learning about aversive stimuli. For example, it is known from fear-conditioning experiments that the amygdala is necessary for learning associations between neutral stimuli and predicted aversive events. Moreover, humans experiencing direct electrical stimulation of the amygdala typically report feelings of fear or anxiety (Gloor et al. 1982; Halgren et al. 1978). Consistent with these findings, monkeys with amygdala lesions fail to respond appropriately to stimuli that generally evoke fear responses, such as snakes (Prather et al. 2001).

However, recent studies suggest that amygdala function is not restricted to aversive events. Neurons in the amygdala signal both positive and negative outcomes associated with odors in rats (Schoenbaum et al. 1998, 1999), and neurons in amygdala are activated when monkeys learn to link neutral cues to either predicted rewards or predicted punishments (Paton et al. 2006). In addition, lesions of the amygdala generally inhibit instrumental learning, the learning of associations between actions and rewards (Balleine et al. 2003; Corbit and Balleine 2005). Furthermore, the amygdala plays a role in social processing, including emotion recognition (e.g., Adolphs et al. 1994; Tremblay and Schultz 1999) and sexual behavior (Aharon et al. 2001); and a recent neuroimaging study suggests that the amygdala is more strongly activated by both attractive and unattractive faces more than by neutral faces, suggesting that the area responds to salient social stimuli regardless of valence (Winston et al. 2007). It is important to note that the amygdala is composed of several different nuclei, each morphologically distinct and differentially connected to other parts of brain. These distinct regions likely have distinct functional roles that will be disentangled only by further research.

6.2.2 Translating Value into Choices: The Neuroeconomic Approach

Value-related signals in the ventral striatum, amygdala, OFC, and other areas modulate ongoing information processing in cortical areas that intervene between sensation and action, such as prefrontal cortex and parietal cortex. This process has been probed using behavioral paradigms, derived from behavioral economics, in which experimental subjects freely choose between alternatives. The revealed preferences of the subject are used to infer the value attributed to available options (Glimcher and Rustichini 2004).

A neuroeconomic approach was first applied explicitly to the neurobiology of decision making by Platt and Glimcher (1999), who probed the impact of expected value on sensory-motor processing in the lateral intraparietal area (LIP), a region of the brain previously linked to visual attention and oculomotor preparation. First defined by Arnaud and Nichole (1982), *expected value* is defined as the product of reward size and reward likelihood associated with a particular option. In Platt and Glimcher's study, monkeys were cued to shift gaze from a central light to one of two peripheral lights to receive a fruit juice reward. In separate blocks of trials, the expected value of orienting to each light was altered by varying either the size of reward or the probability the monkey would be cued to shift gaze to each of the lights. Platt and Glimcher found that LIP neurons signaled a target's expected value prior to cue onset. In a second experiment, in which monkeys could choose freely between the two targets, both neuronal activity in LIP and choice frequency were correlated with expected target value, suggesting that value modulation of sensory-motor processing biases decisions in favor of actions associated with better payoffs.

Sugrue et al. (2004) extended these observations by probing the dynamics of decision-related activity in LIP using a virtual foraging task. In this experiment, the likelihood of rewards associated with each of two targets fluctuated over time, depending on the monkeys' recent rates of choosing each option. Under these conditions, monkeys tended to match the rate of choosing each target to its relative rate of reinforcement over both short and long time scales. Moreover, the responses of individual LIP neurons to a particular target corresponded to the target's history of relative payoffs, with the greatest weight placed on the most recent trials. Together, these and other studies suggest that behavioral decisions may be computed by scaling neuronal responses to sensory stimuli and motor plans by their expected reward value, thus modifying the likelihood of reaching the threshold for eliciting a specific behavioral response (Gold and Shadlen 2001). Endorsing this notion, Bendiksby and Platt (2006) found that activation of LIP neurons was enhanced simply by increasing the motivation of the animal and that this enhancement was associated with predictable changes in behavioral response time. Collectively, these studies demonstrate the role of cortical neurons in guiding decision-making behavior and thereby illustrate the value of a neuroeconomic approach.

6.3 SOCIAL DECISION MAKING

6.3.1 SOCIAL REWARDS IN PRIMATES

Although outcomes such as food acquisition or mating clearly motivate behavior, abstract goals such as information gathering or social interaction can also motivate approach or orienting behavior in the absence of hedonic experience. Furthermore, for group-living species such as humans and many nonhuman primates, the social environment strongly influences the behavioral context in which they pursue rewards, avoid punishments, evaluate risks, and make decisions. Many decisions are motivated by competitive and cooperative interactions with others in a social group; social position not only constrains our behavioral options but constitutes a major target of goal-directed activity. Recent research has also revealed that cooperative transactions (Rilling et al. 2002) or opportunities to punish traitors (de Quervain et al. 2004) can be as motivating to humans as primary rewards such as food and water. Indeed, the observation that individuals will exert physical effort (Aharon et al. 2001) and forgo nonsocial rewards (Fehr and Gachter 2002) to view and/or interact with others demonstrates that social stimuli have positive reinforcement value.

The adaptive significance of navigating a complex social environment explains why social stimuli—an attractive smiling face, a cooperative transaction, the opportunity to punish a traitor—evoke neural activity in some of the same circuits that process primary rewards. Physical features of the face, for example, provide information about genetic quality and affective attitude, and thus also about the quality and likelihood of potential mating opportunities. In a broad variety of situations, people use information encoded in faces to assess the probability and tenor of potential interactions: whether to expect avoidance, aid, or aggression from encountered individuals (e.g., Winston et al. 2002). Likewise, field studies of nonhuman primate behavior have revealed that monkeys preferentially invest in relationships with dominant individuals (Cheney and Seyfarth 1990). Moreover, male primates often use visual cues to predict female mating receptivity (Hrdy and Whitten 1987). These observations suggest that both human and nonhuman primates maintain valuation functions for specific social and reproductive stimuli that guide behavior. These observations implicate a neural system linking social stimuli, such as faces, to the valuation functions guiding behavioral decision making.

Consistent with such a system, a wealth of data indicates that primates find social stimuli to be intrinsically rewarding, and that some types of social stimuli are more reinforcing than others (Anderson 1998; Emery 2000). Rewarding properties of social stimuli for human and nonhuman primates can be inferred from studies of preferential looking. Human infants attend longer to faces expressing anger or surprise rather than fear (Serrano et al. 1992), showing that these social signals evoke different behavioral responses. Likewise, monkeys spend more time looking at pictures of faces looking toward them than pictures with averted gaze (Keating and Keating 1982), likely because directed gaze implies elevated likelihood of interaction. Monkeys also deploy gaze in such a way as to maximize their informational yield: in free-ranging situations, directing their gaze more often toward higher-ranking than lower-ranking animals (McNelis and Boatright-Horowitz 1998); and when viewing faces, preferentially examining mobile, expressive features such as the eyes and mouth (Keating and Keating 1982; Kyes et al. 1992). Similar social biases in orienting have been demonstrated in the visual behavior of humans (Yarbus 1967).

These observations support the hypothesis that the primate brain has evolved mechanisms for attributing positive reward value to social stimuli and that attention is guided by the reinforcing value of inspecting these stimuli. Deaner, Khera, and Platt (2005) explored this hypothesis quantitatively in the laboratory by developing a new economic task in which male rhesus macaques were given a choice between two visual targets. Orienting to one target yielded fruit juice, but orienting to the other yielded fruit juice and the picture of a familiar monkey. By systematically changing the juice payoffs for each target and the pools of images revealed, the authors were able to estimate the subjective value of different types of social and reproductive stimuli in a liquid currency. Their work revealed that male monkeys will forgo larger juice rewards in order to view female sexual signals or faces of high-ranking males, but require these large rewards to choose the faces of low-ranking males (Figure 6.2a). This "pay-per-view" paradigm could easily be extended to study how social valuation might vary when traded against addictive drugs rather than fluids, thus providing a potential model for understanding the impact of addiction on social behavior.

In contrast to the valuation functions governing initial target choice, the durations of sustained gaze associated with each image class hint at the complexity of social influences of behavioral decision. Specifically, monkeys looked at female sexual signals for longer than they looked at either high-ranking or low-ranking male faces, perhaps reflecting differences in the hedonic qualities of these stimuli

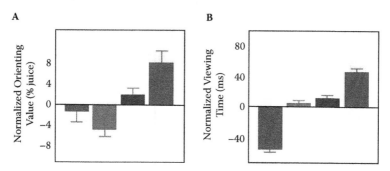

FIGURE 6.2 (Please see color insert following page 112.) Monkeys value social and reproductive information. (a) Mean normalized orienting values for images of familiar monkeys. Orienting values are significantly higher for both the perinea (red bar) and high-status faces (blue bar), in contrast to either the low-status faces (green bar) or gray square (gray bar). Orienting value is measured as the amount of juice required to induce the monkey to choose equally between the image and nonimage target. Therefore, positive numbers reflect the amount of juice the subject monkey will "pay" to see the image; negative numbers reflect the amount of juice required to "bribe" the monkey to see the image. (b) Normalized-looking times for various image classes. Although the monkeys choose to orient more frequently to the high-status faces than the low-status faces, the lengths of time they gaze at both of these image classes are shorter than the time they spend viewing the perinea. Adapted from Deaner, R. O., Khera, A. V., and Platt, M. L. 2005 Monkeys pay per view: adaptive valuation of social images by rhesus macaques. *Curr Biol* 15, 543–8. With permission.

(Figure 6.2b). Male monkeys may thus value the opportunity to view the faces of high-ranking males not because they are hedonically pleasing but rather because they are potentially threatening and thus highly relevant for guiding behavior. This implies that information gathering may activate reward machinery even when this process lacks hedonic value. Such a pairing of negative affect with approach mechanisms is not unprecedented, as the emotion of anger likewise generates behavioral approach paired with strongly negative affect (Ohman et al. 2001). It seems strange that behaviors that are strongly motivating are not necessarily hedonically pleasing, and the mechanisms through which such seeming contradictions are implemented in neuronal circuitry remains unknown. Nonetheless, such contradictions may play an especially important role in drug abuse, in which the hedonic value of the "high" may disappear even as the behavioral motivation to indulge increases.

These results indicate that economic approaches provide a solid foundation upon which to build the study of social reward processes, and likewise that the study of social processes may provide insight into drug abuse, risk, and decision making. We have recently shown that humans, like monkeys, will pay more to view pictures of attractive members of the opposite sex than to view pictures of unattractive ones, even when the reward cues are implicit (Hayden et al. 2007). This technique allowed us to estimate the implicit, abstract value of attending to an attractive member of the opposite sex. Specifically, men place a value of around half a cent (U.S.) on the opportunity to view an attractive woman, whereas the value women place on the opportunity to view attractive men is not different from zero (Figure 6.3). We also showed that the opportunity to view members of the opposite sex is discounted temporally and is exchangeable for effort. These findings suggest that brain processes

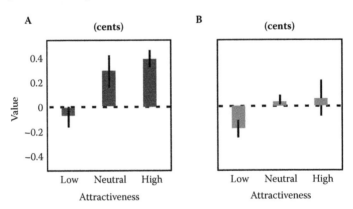

FIGURE 6.3 People implicitly substitute images of the opposite sex for money. (a) Average PSEs (point of subjective equivalence) for all males, expressed in terms of equivalent monetary price (U.S. cents). Males valued the opportunity to view attractive and neutral females, but required a small (nonsignificant) payment to view unattractive females. Bars indicate 1 s.e., estimated by jackknife. (b) Average PSEs for all females in population. Females did not significantly value the opportunity to view attractive and neutral males, and required a payment to view unattractive males. Hayden, B. Y., Parikh, P. C., Deaner, R. O., and Platt, M. L. 2007 Economic principles motivating social attention in humans. *Proc Biol Sci* 274, 1751–1756. With permission.

for evaluating and choosing monetary options are part of a more general system that mediates decisions about other rewards, including social ones.

While the exact mechanisms that relate social and nonsocial reward evaluation remain obscure, their interactions are likely to influence a wide variety of behaviors. Social context strongly patterns behavior in modern society, sometimes promoting and sometimes censuring activities that put individuals—and sometimes bystanders—at increased risk. Examples include peer pressure to engage—or to *not* engage—in activities such as consuming alcohol socially, smoking in public places, wearing seatbelts or motorcycle helmets, participating in sports or dares, and so forth.

6.3.2 Neural Circuits Processing Social Rewards and Punishments

Most research on the neural representation of value and reward in the nonhuman primate brain has utilized primary reinforcers—such as food or juice, or drugs of abuse such as cocaine—which are known to evoke DA responses directly. However, other types of stimuli are likely to play important roles in reward and the development of drug dependence and relapse. Initial drug use typically occurs in a social context, and social environments associated with drug use are potent triggers of relapse. Likewise, social engagement, including both support structures and responsibilities, can protect against development of addiction (Barrett and Turner 2006). Thus, neural processing of social context likely plays an important role in the development of drug dependency. Since most nonhuman primates live in large, structured social groups characterized by social hierarchies and by extensive aggressive and affiliative interaction, it seems likely that the sensitivity of human decisions to social context reflects the evolved specialization of primate reward systems for appropriately motivating and regulating behavior in complex social groups (Insel 2003).

In humans, a varied assortment of positive experiences, including eating chocolate, hearing pleasant music, or reading a funny cartoon, evokes activity in reward-related brain regions including OFC and nucleus accumbens (NAcc). Most single-neuron studies of this region in rodent (Schoenbaum et al. 1998) and monkey (Rolls and Baylis 1994; Tremblay and Schultz 1999) models have focused on the role of the OFC in association formation in general, rather than social processing in particular. However, lesion (Butter and Snyder 1972; Hornak et al. 2003; Myers et al. 1973) and functional magnetic resonance imaging (fMRI) (de Quervain et al. 2004; O'Doherty et al. 2003; Rilling et al. 2002) studies have both underscored the importance of OFC for social interaction and decision making. Anatomically, OFC is connected with the amygdala and inferotemporal cortex, and interactions between these three regions appear to play a critical role in social processing. For example, intact amygdala (Adolphs et al. 1994) and inferotemporal cortex (Ellis et al. 1989) are required for successful decoding of facial expressions and face identification, respectively.

Like the midbrain and prefrontal lobes, the parietal cortex plays an important role in social decisions, just as it does in nonsocial ones. Motivational states related to reward expectation systematically modulate the activity of parietal neurons (Bendiksby and Platt 2006; Platt and Glimcher 1999), and recent studies in our lab (Klein et al. 2008) indicate that these neurons encode the reward value of a social stimulus and fluid rewards similarly when contrasted using a pay-per-view paradigm. These data indicate that neural circuits controlling visual orienting use integrated social and nonsocial reward signals to determine the behavioral salience of visual stimuli. We speculate that these signals arise within the brain reward circuitry described above, specifically the ventral striatum and OFC.

6.4 RISK AND AMBIGUITY

6.4.1 Risk and Decision Making

Early ethological models of behavior assumed, for simplicity, that animals maintained complete knowledge of the environment and that reward contingencies were deterministic. In practice, however, uncertainty about environmental contingencies places strong constraints on behavior. The impact of uncertainty on choice has long been acknowledged in economics, which defines the spread of an outcome's known probability as risk. In the 18th century, Daniel Bernoulli proposed that the subjective (i.e., as determined by the economic agent) utility of a sum of money is a concave function of its face value (Figure 6.4). It follows directly from this idea that decision makers will be generally risk averse for potential gains. The generality of risk aversion has been widely observed for both humans and animals of several species in several contexts. More recent studies have fleshed out this picture of animal behavior under risk. For example, it has been established that when reward sizes are held constant but the delay until reward is unpredictable, animals will generally prefer the risky option (Bateson and Kacelnik 1997). This behavior may reflect the well-known hyperbolic discounting of delays (Mazur 1987), which effectively leads to a convex value function and thus promotes risk seeking. Interestingly, addicts to opioids

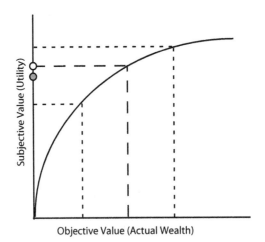

FIGURE 6.4 Concave subjective utility curve. The subjective utility of a single fixed value (white dot, dashed line) is greater than the average utility from two variable values with the same mean objective value (gray dot, dotted line). This indicates an aversion to risk.

(Madden et al. 1997) and nicotine (Bickel et al. 1999) discount future rewards more steeply than do nonaddicted controls, and opioid-dependent subjects discount heroin more steeply than money. It is unknown whether this difference in temporal discounting is actually caused by the addiction itself or it reflects a trait that occurs in those who are predisposed to addiction.

In a seminal study, Caraco and colleagues (1980) observed that risk preferences of yellow-eyed juncos depend on ambient temperature. These small birds were given the option of choosing a tray with a fixed number of millet seeds or a tray with a probabilistically varying number of seeds with the same mean as the fixed option. At 19°C juncos preferred the fixed option, but at 1°C they preferred the variable option. The proposed explanation for this switch from risk aversion to risk seeking is that at the higher temperature the rate of gain from the fixed option was sufficient to maintain the bird on a positive energy budget. At the lower temperature, however, energy expenditures were elevated, so the fixed option was no longer adequate to meet the animal's energy needs. When cold, the bird's best chance for survival was to gamble on the risky option since it might yield a higher rate of return than the fixed option. In other words, at low temperatures, the bird's utility function switched from concave to convex, and the possibility of the larger reward is enough to bias the bird to risk seeking. This experiment has a direct correlate in the addiction literature: Preference for a variable drug source over a fixed drug source was induced in drug addicts when they read scripts that simulated drug withdrawal, whereas this preference was reversed when the subjects read scripts that simulate drug satiation (Bickel et al. 2004).

Caraco's observations strongly suggest that sensitivity to risk is an evolutionarily ancient neural adaptation supporting adaptive decision making. The foregoing discussion makes plain that the discrepancy between observable outcomes and subjective preferences in decision making under risk offers a powerful paradigm for investigating the neural mechanisms underlying adaptive decision making. In particular, animals operating under stress and uncertainty are more inclined to choose riskier behaviors, presumably because the status quo presents a fairly bleak outlook for evolutionary success. The fact that economics describes common elements of behavior in animals "seeking" to maximize genetic fitness as well as in humans seeking to maximize monetary wealth suggests a common underlying mechanism.

6.4.2   Neural Basis of Decision Making under Conditions of Risk

Since individuals often demonstrate strong attraction or aversion to uncertain options, risk sensitivity provides a powerful assay to dissociate subjective preferences from objective rewards. One important neurobiological question is where and how the brain creates subjective preferences incorporating outcome uncertainty. Oculomotor choice tasks have been extensively employed to explore this question. As mentioned previously, LIP serves as a critical site for integrating information about rewards and using this information to generate oculomotor plans (Platt and Glimcher 1999; Sugrue et al. 2004). However, it remains unclear which of its diverse input structures provides the relevant information about rewards. Based on anatomical considerations, one likely candidate is the posterior cingulate cortex (CGp). Indeed, CGp, among other limbic areas, appears to evaluate reward outcomes that depend on the animal's subjective state and then inform brain areas, such as LIP, that translate value signals into action (Dean et al. 2004; McCoy et al. 2003). CGp is interconnected with oculomotor and reward-related brain areas (Vogt et al. 1992), and CGp neurons respond after visual events, suggesting an evaluative role in visual orienting (McCoy et al. 2003; Olson et al. 1996). Moreover, CGp neurons signal omission of predicted rewards (McCoy et al. 2003).

We have used oculomotor tasks to determine whether CGp signals subjective risk preferences when representing targets with highly variable or relatively constant payoff (McCoy and Platt 2005). Like other decision makers, monkeys exhibit reliable risk preferences, choosing a risky option in a visual gambling task even when it pays less, on average, than a safe option (Figure 6.5). CGp neurons closely reflect this behavioral bias, rather than representing the objective expected value of each target. Further, CGp neuronal activity correlated with subjective utility estimated from choices and rewards received (McCoy and Platt 2005). These data indicate that

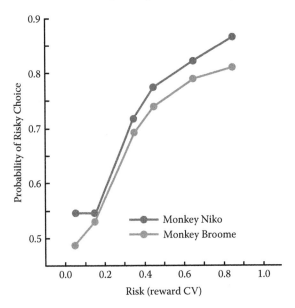

FIGURE 6.5 Macaque monkeys are sensitive to risk and prefer to gamble for a large or small juice reward rather than opt for a sure midsized reward. These two monkeys chose the risky target more than half the time, and the probability of choosing the risky target increased with increasing risk (logistic regression coefficients: Broome, 2.442, $p < 0.0000001$; Niko, 2.426, $p < 0.0000001$). CV = coefficient of variation of rewards associated with the risky target, a dimensionless measure of relative risk. Adapted from McCoy, A. N., and Platt, M. L. 2005 Risk-sensitive neurons in macaque posterior cingulate cortex. *Nat Neurosci* 8, 1220–1227. With permission.

CGp plays a critical role in gambling decisions by binding subjective reward values to potential targets of action.

Why do subjective and objective measures of value diverge? One possibility is that monkeys prefer the risky option because they focus on the large reward and ignore bad outcomes—a possibility suggested by prior behavioral studies (Rachlin et al. 2000; Tversky and Kahneman 1981) heretofore unexplored in neurobiological studies of reward and decision making. We tested this hypothesis directly by examining the relationship between risk preference and delay between trials (Hayden et al. 2007). We found that preference for risk declines with increasing delays and reverses when delays increase beyond 45 seconds. These findings are explained by "string theory" (Rachlin 2000), which proposes that the salience of the large reward and the expected delay until that reward can be obtained influence the valuation of a risky option. Interestingly, similar processes have been proposed to operate in humans who pursue the immediate, intense "high" of certain drugs of abuse, while simultaneously discounting the delayed, longer-term "low" of withdrawal (Bernheim and Rangel 2004; Bickel et al. 1999).

Further support for the importance of salience in decision under risk comes from the finding that monkeys discriminate small changes in the size of a large reward but do not discriminate equivalent changes in the size of the small reward (Hayden et al. under review). Consistent with these behavioral data, CGp neurons discriminate large rewards from medium rewards but do not distinguish small and medium rewards in the gambling task. Moreover, neural responses precisely predict the sensitivity of the monkey to wins and losses in a gamble. These results are consistent with the idea that neural signals in CGp are subjectively anchored to the jackpot—an idea that echoes the availability heuristic in economics (Tversky and Kahneman 1973).

Neuroimaging studies in humans have revealed that preference for a risky option is associated with increases in neuronal activity in the ventral striatum and posterior parietal cortex (Kuhnen and Knutson 2005; Preuschoff et al. 2006; Tom et al. 2007). Moreover, choosing a risky option activates the dorsal striatum and precuneus (Rustichini et al. 2005). Building on these studies, economists often distinguish between "risky" options, with known probabilities, and "ambiguous" options, in which the outcome probabilities are not known. The neural mechanisms mediating decision under risk or ambiguity were explored by Hsu and colleagues as well as Huettel and colleagues (Hsu et al. 2005; Huettel et al. 2006). In the latter study, Huettel and colleagues probed choices between monetary gambles while human subjects were scanned using fMRI. They found that subjects' preferences for ambiguity were correlated with activation in prefrontal cortex and that preferences for risk were correlated with activity in parietal cortex (Figure 6.6). This study was the first direct parametric link of economic preferences to activation of specific brain regions.

These results highlight potential targets for electrophysiological studies in monkeys and also provide a paradigm for investigating the influence of candidate genes that may contribute to structural and functional differences between individuals showing behavioral differences in risk sensitivity and preference.

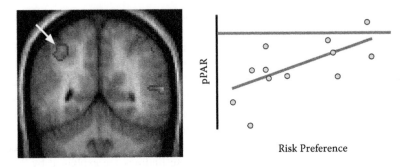

FIGURE 6.6 (Please see color insert following page 112.) Subjective economic preferences predict changes in brain activation. A correlation exists between cortical activity in the posterior parietal cortex (see left, white arrow) and behavioral preference for risk. Right, each subject's mean preference value and fMRI parameter value (a measure of relative brain activity) is indicated by a single circle. Correlation = −0.62, $p < 0.00001$. Adapted from Huettel, S. A., Stowe, C. J., Gordon, E. M., Warner, B. T., and Platt, M. L. 2006. Neural signatures of economic preferences for risk and ambiguity. *Neuron* 49, 765–775. With permission.

6.5   ADDICTION AND THE NEUROBIOLOGY OF DECISION MAKING

6.5.1   DRUGS, DECISIONS, AND DOPAMINE

Physiologically, the most completely characterized mechanism for establishing drug dependence involves the mesolimbic dopamine "reward" system (Hyman and Malenka 2001; Wise 1987, 1998). Neurons within the ventral midbrain release dopamine into the ventral striatum, specifically the NAcc, among other places. A vast literature shows that the activity of these dopaminergic cells and the subsequent release of dopamine in the NAcc play an essential role in mediating the hedonic value of primary reinforcers, such as food or juice, as well as the motivation necessary for operant conditioning to stimuli predictive of such primary reinforcers (Berridge and Robinson 1998; Cromwell et al. 2005; McClure et al. 2003; Phillips et al. 2003; Roitman et al. 2004; Schultz et al. 1992; Wise 2004). Many drugs with a high potential for creating dependency co-opt the function of this dopaminergic reward circuitry directly by altering dopamine release and reuptake (cocaine, methamphetamine). Dopamine contributes to the rewarding properties of stimulants such as methamphetamine (Schultz 2001). Long-term methamphetamine abuse largely results in damage to dopaminergic nerve terminals and diminished dopamine release in reward areas of the brain, particularly the NAcc, and is associated with pathologies in social behavior such as paranoia, anxiety, and violent behavior (Iyo et al. 2004). Other drugs with high potential for creating dependency are believed to alter the function of the mesolimbic dopamine systems indirectly, through interactions with other neurotransmitter systems (Balfour et al. 2000; Heinz 2002). For example, an increase in serotonin level enhances the rewarding properties of cocaine, whereas depletion of serotonin has the opposite effect (Neisewander and Acosta 2004; Peltier and Schenk 1993).

Intriguingly, dopamine has also been shown to contribute to social reward and motivation. The best evidence for the role of DA in social reward processing comes from studies of pair-bonding in voles, small rodents in which different species mate either monogamously or nonmonogamously (Insel and Young 2000, 2001). In monogamous voles males and females form strong pair-bonds after mating, whereas in nonmonogamous voles such pair bonds are not formed. Pair-bonding depends crucially on the action of dopamine within reward circuitry, including NAcc, mPFC, and OFC. DA is associated with both affiliative and aggressive components of pair-bonding by monogamous males (Keverne and Curley 2004). In NAcc, D_2-type dopamine receptors are involved in affiliative behavior, whereas D_1-type receptors are involved in aggressive behavior. Relatedly, monogamous voles, but not nonmonogamous voles, have high levels of oxytocin in the NAcc and arginine-vasopression in the ventral pallidum, ventral forebrain, and mesolimbic DA reward pathway. DA receptors are also localized within the mPFC and OFC, and these receptors appear to interact with oxytocin and vasopressin receptors in pair-bonding (Smeltzer et al. 2006). Thus, work on rodents, as well as sheep (Kendrick 2004), indicates that DA signaling in the NAcc, and possibly OFC, is critical for social motivation and pair-bonding. Several studies suggest that DA also contributes specifically to social reward processing in primates and humans, where social bonds are more complex. Seeing an image of a loved one (Aron et al. 2005) or an attractive member of the opposite sex (Aharon et al. 2001; Kampe et al. 2001) activates dopaminergic pathways in humans. Moreover, dysfunction of these circuits can be associated with autism (Kendrick 2004) as well as social anxiety disorder (Schneier et al. 2002). Precisely how this circuitry functions in humans remains elusive, but it is clear that just as social context can influence the likelihood of risky behaviors including drug use, so too can DA-targeting drugs strongly alter the mechanisms through which humans relate to one another. It is important to note that such drugs include not only drugs of abuse but also legal prescription drugs such as Ritalin.

The dopaminergic system may also play a crucial role in decision making in uncertain or risky contexts. Recordings in DA neurons following presentation of cues that provide information about the likelihood of an uncertain reward indicate that their phasic responses encode the likelihood of receiving a reward (Fiorillo et al. 2003). Interestingly, their tonic responses following the presentation of the cue are correlated with the amount of uncertainty about the reward: Responses are maximal for a cue that is 50% valid and decrease when there is more certainty. Rapid phenylalanine/tyrosine depletion (RPTD), which selectively reduces dopamine levels in ventral striatum in humans (Mehta et al. 2005), significantly reduces subjects' motivation to bet money in a simulated gambling task (Roiser et al. 2005).

6.5.2 Drugs, Decisions, and Serotonin

Serotonin (5HT) not only modulates the role of other neurotransmitters with respect to addiction but is itself a target of some drugs of abuse. MDMA ("ecstasy" or "X") is a relatively new drug of abuse that has the potential to create physiological dependence, as demonstrated by studies of self-administration and withdrawal in rodents and primates (Jansen 1999). In these respects, it is similar to cocaine or methamphetamine.

However, there are important pharmacological and user-described subjective differences. While all these drugs alter monoamine neurotransmission, cocaine and methamphetamine exert their primary affects on dopamine, while the primary target of MDMA is the serotonin transporter (McCann et al. 2005). Subjective reports of recreational use of these drugs largely overlap. However, one unique effect of MDMA ubiquitously reported is a feeling of "openness" and "empathy" toward other people. Following short-term MDMA-induced enhancement in serotonin transmission, serotonin levels in the brain become depleted, and this depletion is associated with social isolation, social anxiety, and reduced sex drive (Morton 2005). In combination with the efficacy of selective serotonin reuptake inhibitors (SSRIs) in the treatment of social anxiety disorder (Raj 2004), these observations suggest an important role for serotonin in social reward processing.

Serotonin has also been implicated in social behavior across diverse clades of animals from arthropods to mammals (Insel and Winslow 1998). A growing body of evidence supports a role for 5HT in the mediation of behavioral inhibition as well as responses to aversive stimuli, both of which are relevant to social functioning (Daw et al. 2002). Increased 5HT function enhances social affiliation (Insel and Winslow 1998; Knutson et al. 1998; Raleigh et al. 1991), whereas reductions in 5HT signaling have been implicated in social impulsivity and aggression (Raleigh et al. 1991; Westergaard et al. 1999). Applying agonists to the 5HT-1b receptor appears to inhibit the display of aggressive behavior in mice, and social impulsivity is inversely related to the degree of serotonin metabolite present in nonhuman primates (Fairbanks et al. 2001). 5HT levels are also related to social rank in primates, a fact that may be related to each animal's ability to inhibit antisocial behaviors such as aggression (Raleigh et al. 1991; Westergaard et al. 1999). A decrease in serotonin levels through tryptophan depletion increases aggressiveness in monkeys (Bjork et al. 1999; Edwards and Kravitz 1997), and in humans it decreases the likelihood that one will cooperate with a partner while playing the Prisoner's Dilemma game (Wood et al. 2006).

With regard to drug use, hallucinogens targeting the 5HT system can dramatically alter the senses of self, of attachment, and of social importance. Likewise, the club drug MDMA enhances DA and especially 5HT release and is associated with feelings of social empathy and well-being. Long-term or high-dose use of MDMA, however, can damage serotonergic neurons, and even single doses can produce serotonin depletion, associated with depressed mood, social isolation, and reduced sex drive. As is the case with DA, however, it is important to recognize that various legal drugs may also have psychosocial consequences as result of their alteration of the 5HT system. In particular, commonly prescribed antidepressants, including Prozac, Zoloft, and other SSRIs, alter the 5HT system. Sexual side effects are common—for example, altered sexual desire, impaired performance, or anorgasmia (Clayton and Montejo 2006). Of even greater concern, SSRIs may be associated with increased impulsive or suicidal behaviors in depressed children (Mann et al. 2005).

Evidence also links serotonin to decision making in risky situations, perhaps because this neurotransmitter modulates the ability to discriminate subtle variations in gains and losses. For example, Rogers and colleagues (2003) found that rapid tryptophan depletion (RTD), which reduces serotonin levels in the central nervous system, influenced the pattern of decisions individuals made in a risky decision-making task. Subjects chose between gambles that differed in reward and loss magnitudes and outcome probabilities (Rogers et al. 2003). The authors found that RTD reduced subjects' ability to discriminate between options with different expected gains (i.e., reduced sensitivity to expected value of gains), although no such effects were observed for losses or probabilities. On the other hand, a polymorphism that determines the transcription efficiency of the serotonin transporter has been associated with amygdala activity, fear conditioning, trait anxiety, and susceptibility to depression, which suggests that serotonin has a role in regulating sensitivity to aversive stimuli (Brown and Hariri 2006; Caspi et al. 2003).

Adaptive decision making in contexts associated with social complexity or outcome uncertainty thus requires the seamless integration of serotonin- and dopamine-signaling systems within the brain, so it is not surprising that disruptions in either one can impair decision making, with important implications for understanding addiction. Indeed, the effects of dopamine and serotonin dysfunction on decision making are often the most severe when choosing between risky options or in social situations. Such impairments may reflect disruption of the normal antagonistic influences of dopamine and serotonin on reward processing by neurons in the prefrontal cortex and basal ganglia (Daw et al. 2002). Understanding the neurobiology of addiction may therefore be advanced by considering the effects of outcome uncertainty and social context on decision making, an endeavor we believe will be facilitated by the use of neuroeconomic techniques (cf. Bernheim and Rangel 2004).

6.6   CONCLUSIONS

Ultimately, the nervous systems of humans and other animals evolved to promote behaviors that enhance biological fitness, such as acquiring food and shelter, attracting mates, avoiding predators, and prevailing over competitors. In this view, the nervous system of any animal constitutes a suite of morphological and behavioral adaptations for surmounting specific environmental and social challenges. Drug addiction is a disease that hijacks valuation systems in the brain, resulting in behavioral patterns that no longer optimize biological fitness. Because human beings are such an intensely social species, widespread drug abuse not only takes a toll on the physical and psychological well-being of the individuals involved but has severe repercussions for the community as a whole. Although still in the early stages, the union of ethology, economics, and neuroscience—the emerging field of neuroeconomics—offers a potentially powerful way to study the neural mechanisms underlying reward, punishment, decision, and the pharmacological and psychological processes that underlie risk taking and drug addiction. In particular, we find that the neural processes involved in processing social context, and in evaluating risk and uncertainty, are particularly rich in insights and testable predictions for human patterns of drug use.

REFERENCES

Adolphs, R., Tranel, D., Damasio, H., and Damasio, A. 1994 Impaired recognition of emotion in facial expressions following bilateral damage to the human amygdala. *Nature* 372, 669–672.

Aharon, I., Etcoff, N., Ariely, D., Chabris, C. F., O'Connor, E., and Breiter, H. C. 2001 Beautiful faces have variable reward value: fMRI and behavioral evidence. *Neuron* 32, 537–551.

Anderson, J. R. 1998 Social stimuli and social rewards in primate learning and cognition. *Behavioural Processes* 42, 159–175.

Arnauld, A., and Nichole, P. 1982 *The art of thinking: Port-Royal logic*. Indianapolis, IN: Bobbs-Merrill.

Aron, A., Fisher, H., Mashek, D. J., Strong, G., Li, H., and Brown, L. L. 2005 Reward, motivation, and emotion systems associated with early-stage intense romantic love. *J Neurophysiol* 94, 327–337.

Balfour, D. J., Wright, A. E., Benwell, M. E., and Birrell, C. E. 2000 The putative role of extra-synaptic mesolimbic dopamine in the neurobiology of nicotine dependence. *Behav Brain Res* 113, 73–83.

Balleine, B. W., Killcross, A. S., and Dickinson, A. 2003 The effect of lesions of the basolateral amygdala on instrumental conditioning. *Journal of Neuroscience* 23, 666–675.

Barrett, A. E., and Turner, J. 2006 Family structure and substance use problems in adolescence and early adulthood: examining explanations for the relationship. *Addiction* 101, 109–120.

Bateson, M., and Kacelnik, A. 1997 Starlings' preferences for predictable and unpredictable delays to food. *Animal Behaviour* 53, 1129–1142.

Bendiksby, M. S., and Platt, M. L. 2006 Neural correlates of reward and attention in macaque area LIP. *Neuropsychologia* 44, 2411–2420.

Bernheim, B. D., and Rangel, A. 2004 Addiction and cue-triggered decision processes. *American Economic Review* 94, 1558–1590.

Berridge, K. C., and Robinson, T. E. 1998 What is the role of dopamine in reward: hedonic impact, reward learning, or incentive salience? *Brain Res Brain Res Rev* 28, 309–369.

Bickel, W. K., Giordano, L. A., and Badger, G. J. 2004 Risk-sensitive foraging theory elucidates risky choices made by heroin addicts. *Addiction* 99, 855–861.

Bickel, W. K., Odum, A. L., and Madden, G. J. 1999 Impulsivity and cigarette smoking: delay discounting in current, never, and ex-smokers. *Psychopharmacology* 146, 447–454.

Bjork, J. M., Dougherty, D. M., Moeller, F. G., Cherek, D. R., and Swann, A. C. 1999 The effects of tryptophan depletion and loading on laboratory aggression in men: time course and a food-restricted control. *Psychopharmacology (Berl)* 142, 24–30.

Brown, S. M., and Hariri, A. R. 2006 Neuroimaging studies of serotonin gene polymorphisms: exploring the interplay of genes, brain, and behavior. *Cogn Affect Behav Neurosci* 6, 44–52.

Butter, C. M., and Snyder, D. R. 1972 Alterations in aversive and aggressive behaviors following orbital frontal lesions in rhesus monkeys. *Acta Neurobiol Exp (Wars)* 32, 525–565.

Caraco, T., Martindale, S., and Whittam, T. S. 1980 An empirical demonstration of risk-sensitive foraging preferences. *Anim Behav* 28, 820–830.

Caspi, A., Sugden, K., Moffitt, T. E., Taylor, A., Craig, I. W., Harrington, H., McClay, J., Mill, J., Martin, J., Braithwaite, A., and Poulton, R. 2003 Influence of life stress on depression: moderation by a polymorphism in the 5-HTT gene. *Science* 301, 386–389.

Cheney, D. L., and Seyfarth, R. M. 1990 *How monkeys see the world: Inside the mind of another species*. Chicago: University of Chicago Press.

Clayton, A. H., and Montejo, A. L. 2006 Major depressive disorder, antidepressants, and sexual dysfunction. *J Clin Psychiatry* 67, 33–37.

Corbit, L. H., and Balleine, B. W. 2005 Double dissociation of basolateral and central amygdala lesions on the general and outcome-specific forms of pavlovian-instrumental transfer. *J Neurosci* 25, 962–970.

Critchley, H. D., and Rolls, E. T. 1996 Hunger and satiety modify the responses of olfactory and visual neurons in the primate orbitofrontal cortex. *J Neurophysiol* 75, 1673–1686.

Cromwell, H. C., Hassani, O. K., and Schultz, W. 2005 Relative reward processing in primate striatum. *Exp Brain Res* 162, 520–525.

Daw, N. D., Kakade, S., and Dayan, P. 2002 Opponent interactions between serotonin and dopamine. *Neural Netw* 15, 603–616.

de Quervain, D. J., Fischbacher, U., Treyer, V., Schellhammer, M., Schnyder, U., Buck, A., and Fehr, E. 2004 The neural basis of altruistic punishment. *Science* 305, 1254–1258.

Dean, H. L., Crowley, J. C., and Platt, M. L. 2004 Visual and saccade-related activity in macaque posterior cingulate cortex. *J Neurophysiol* 92, 3056–3068.

Deaner, R. O., Khera, A. V., and Platt, M. L. 2005 Monkeys pay per view: adaptive valuation of social images by rhesus macaques. *Curr Biol* 15, 543–8.

Edwards, D. H., and Kravitz, E. A. 1997 Serotonin, social status and aggression. *Curr Opin Neurobiol* 7, 812–819.

Ellis, A. W., Young, A. W., and Critchley, E. M. R. 1989 Loss of memory for people following temporal-lobe damage. *Brain* 112, 1469–1483.

Emery, N. J. 2000 The eyes have it: the neuroethology, function and evolution of social gaze. *Neurosci Biobehav Rev* 24, 581–604.

Fairbanks, L. A., Melega, W. P., Jorgensen, M. J., Kaplan, J. R., and McGuire, M. T. 2001 Social impulsivity inversely associated with CSF 5-HIAA and fluoxetine exposure in vervet monkeys. *Neuropsychopharmacology* 24, 370–378.

Fehr, E., and Gachter, S. 2002 Altruistic punishment in humans. *Nature* 415, 137–140.

Fiorillo, C. D., Tobler, P. N., and Schultz, W. 2003 Discrete coding of reward probability and uncertainty by dopamine neurons. *Science* 299, 1898–1902.

Glimcher, P. W., and Rustichini, A. 2004 Neuroeconomics: the consilience of brain and decision. *Science* 306, 447–452.

Gloor, P., Olivier, A., Quesney, L. F., Andermann, F., and Horowitz, S. 1982 The role of the limbic system in experiential phenomena of temporal-lobe epilepsy. *Ann Neurol* 12, 129–144.

Gold, J. I., and Shadlen, M. N. 2001 Neural computations that underlie decisions about sensory stimuli. *Trends Cogn Sci* 5, 10–16.

Gottfried, J. A., O'Doherty, J., and Dolan, R. J. 2002 Appetitive and aversive olfactory learning in humans studied using event-related functional magnetic resonance imaging. *J Neurosci* 22, 10829–10837.

Halgren, E., Walter, R. D., Cherlow, D. G., and Crandall, P. H. 1978 Mental phenomena evoked by electrical-stimulation of human hippocampal formation and amygdala. *Brain* 101, 83–117.

Hayden, B. Y., Parikh, P. C., Deaner, R. O., and Platt, M. L. 2007 Economic principles motivating social attention in humans. *Proc Biol Sci* 274, 1751–1756.

Heinz, A. 2002 Dopaminergic dysfunction in alcoholism and schizophrenia—psychopathological and behavioral correlates. *Eur Psychiatry* 17, 9–16.

Hornak, J., Bramham, J., Rolls, E. T., Morris, R. G., O'Doherty, J., Bullock, P. R., and Polkey, C. E. 2003 Changes in emotion after circumscribed surgical lesions of the orbitofrontal and cingulate cortices. *Brain* 126, 1691–712.

Hosokawa, T., Kato, K., Inoue, M., and Mikami, A. 2007 Neurons in the macaque orbitofrontal cortex code relative preference of both rewarding and aversive outcomes. *Neurosci Res* 57, 434–445.

Hrdy, S. B., and Whitten, P. L. 1987 Patterning of sexual activity. In *Primate societies* (ed. D.L. Cheney, B.B. Smuts, R.M. Seyfarth, R.W. Wrangham and T.T. Struhsaker). Chicago: University of Chicago Press.

Hsu, M., Bhatt, M., Adolphs, R., Tranel, D., and Camerer, C. F. 2005 Neural systems responding to degrees of uncertainty in human decision-making. *Science* 310, 1680–1683.

Huettel, S. A., Stowe, C. J., Gordon, E. M., Warner, B. T., and Platt, M. L. 2006 Neural signatures of economic preferences for risk and ambiguity. *Neuron* 49, 765–775.

Hyman, S. E., and Malenka, R. C. 2001 Addiction and the brain: the neurobiology of compulsion and its persistence. *Nat Rev Neurosci* 2, 695–703.

Insel, T. R. 2003 Is social attachment an addictive disorder? *Physiol Behav* 79, 351–357.

Insel, T. R., and Winslow, J. T. 1998 Serotonin and neuropeptides in affiliative behaviors. *Biol Psychiatry* 44, 207–219.

Insel, T. R., and Young, L. J. 2000 Neuropeptides and the evolution of social behavior. *Curr Opin Neurobiol* 10, 784–789.

Insel, T. R., and Young, L. J. 2001 The neurobiology of attachment. *Nat Rev Neurosci* 2, 129–136.

Iyo, M., Sekine, Y., and Mori, N. 2004 Neuromechanism of developing methamphetamine psychosis: A neuroimaging study. In *Current status of drug dependence / abuse studies: cellular and molecular mechanisms of drugs of abuse and neurotoxicity*, vol. 1025, pp. 288–295.

Jansen, K. L. R. 1999 Ecstasy (MDMA) dependence. *Drug Alcohol Depend* 53, 121–124.

Kampe, K. K., Frith, C. D., Dolan, R. J., and Frith, U. 2001 Reward value of attractiveness and gaze. *Nature* 413, 589.

Keating, C. F., and Keating, E. G. 1982 Visual scan patterns of rhesus monkeys viewing faces. *Perception* 11, 211–219.

Kendrick, K. M. 2004 The neurobiology of social bonds. *J Neuroendocrinol* 16, 1007–1008.

Keverne, E. B., and Curley, J. P. 2004 Vasopressin, oxytocin and social behaviour. *Curr Opin Neurobiol* 14, 777–783.

Klein, J. T., R. O. Deaner, et al. (2008). Neural correlates of social target value in macaque parietal cortex. *Curr Biol* 18(6): 419–424.

Knutson, B., Wolkowitz, O. M., Cole, S. W., Chan, T., Moore, E. A., Johnson, R. C., Terpstra, J., Turner, R. A., and Reus, V. I. 1998 Selective alteration of personality and social behavior by serotonergic intervention. *Am J Psychiatry* 155, 373–379.

Kringelbach, M. L., O'Doherty, J., Rolls, E. T., and Andrews, C. 2003 Activation of the human orbitofrontal cortex to a liquid food stimulus is correlated with its subjective pleasantness. *Cereb Cortex* 13, 1064–1071.

Kuhnen, C. M., and Knutson, B. 2005 The neural basis of financial risk taking. *Neuron* 47, 763–770.

Kyes, R. C., Mayer, K. E., and Bunnell, B. N. 1992 Perception of stimuli presented as photographic slides in cynomolgus macaques (*Macaca fascicularis*). *Primates* 33, 407–412.

Madden, G. J., Petry, N. M., Badger, G. J., and Bickel, W. K. 1997 Impulsive and self-control choices in opioid-dependent patients and non-drug-using control participants: drug and monetary rewards. *Exp Clin Psychopharmacol* 5, 256–262.

Mann, J. J., Emslie, G., Baldessarini, R. J., Beardslee, W., Fawcett, J. A., Goodwin, F. K., Leon, A. C., Meltzer, H. Y., Ryan, N. D., Shaffer, D., and Wagner, K. D. 2005 ACNP Task Force Report on SSRIs and suicidal behavior in youth. *Neuropsychopharmacology* 31, 473–492.

Mazur, J. 1987 An adjusting procedure for studying delayed reinforcement: the effect of delay and intervening events, vol. V (ed. M. Commons, J. Mazur, J. Nevin and H. Rachlin), pp. 55–73. London: Erlbaum.

McCann, U. D., Szabo, Z., Seckin, E., Rosenblatt, P., Mathews, W. B., Ravert, H. T., Dannals, R. F., and Ricaurte, G. A. 2005 Quantitative PET studies of the serotonin transporter in MDMA users and controls using [11C]McN5652 and [11C]DASB. *Neuropsychopharmacology* 30, 1741–1750.

McClure, S. M., Berns, G. S., and Montague, P. R. 2003 Temporal prediction errors in a passive learning task activate human striatum. *Neuron* 38, 339–346.

McCoy, A. N., Crowley, J. C., Haghighian, G., Dean, H. L., and Platt, M. L. 2003 Saccade reward signals in posterior cingulate cortex. *Neuron* 40, 1031–1040.

McCoy, A. N., and Platt, M. L. 2005 Risk-sensitive neurons in macaque posterior cingulate cortex. *Nat Neurosci* 8, 1220–1227.

McNelis, N. L., and Boatright-Horowitz, S. L. 1998 Social monitoring in a primate group: the relationship between visual attention and hierarchical ranks. *Animal Cognition* 1, 65–69.

Mehta, M. A., Gumaste, D., Montgomery, A. J., McTavish, S. F., and Grasby, P. M. 2005 The effects of acute tyrosine and phenylalanine depletion on spatial working memory and planning in healthy volunteers are predicted by changes in striatal dopamine levels. *Psychopharmacology (Berl)* 180, 654–663.

Montague, P. R., and Berns, G. S. 2002 Neural economics and the biological substrates of valuation. *Neuron* 36, 265–284.

Morton, J. 2005 Ecstasy: pharmacology and neurotoxicity. *Curr Opin Pharmacol* 5, 79–86.

Myers, R. E., Swett, C., and Miller, M. 1973 Loss of social group affinity following prefrontal lesions in free-ranging macaques. *Brain Res* 64, 257–269.

Neisewander, J. L., and Acosta, J. I. 2004 Serotonin systems modulate cocaine-seeking behavior: implications for cocaine craving and relapse. *Neuropsychopharmacology* 29, S51–S52.

O'Doherty, J., Rolls, E. T., Francis, S., Bowtell, R., and McGlone, F. 2001 Representation of pleasant and aversive taste in the human brain. *J Neurophysiol* 85, 1315–1321.

O'Doherty, J., Winston, J., Critchley, H., Perrett, D., Burt, D. M., and Dolan, R. J. 2003 Beauty in a smile: the role of medial orbitofrontal cortex in facial attractiveness. *Neuropsychologia* 41, 147–155.

Ohman, A., Lundqvist, D., and Esteves, F. 2001 The face in the crowd revisited: a threat advantage with schematic stimuli. *J Pers Soc Psychol* 80, 381–396.

Olds, J., and Milner, P. 1954 Positive reinforcement produced by electrical stimulation of septal area and other regions of rat brain. *J Comp Physiol Psychol* 47, 419–427.

Olson, C. R., Musil, S. Y., and Goldberg, M. E. 1996 Single neurons in posterior cingulate cortex of behaving macaque: eye movement signals. *J Neurophysiol* 76, 3285–3300.

Padoa-Schioppa, C., and Assad, J. A. 2006 Neurons in the orbitofrontal cortex encode economic value. *Nature* 441, 223–226.

Paton, J. J., Belova, M. A., Morrison, S. E., and Salzman, C. D. 2006 The primate amygdala represents the positive and negative value of visual stimuli during learning. *Nature* 439, 865–870.

Peltier, R., and Schenk, S. 1993 Effects of serotonergic manipulations on cocaine self-administration in rats. *Psychopharmacology* 110, 390–394.

Phillips, P. E., Stuber, G. D., Heien, M. L., Wightman, R. M., and Carelli, R. M. 2003 Subsecond dopamine release promotes cocaine seeking. *Nature* 422, 614–618.

Platt, M. L., and Glimcher, P. W. 1999 Neural correlates of decision variables in parietal cortex. *Nature* 400, 233–238.

Prather, M. D., Lavenex, P., Mauldin-Jourdain, M. L., Mason, W. A., Capitanio, J. P., Mendoza, S. P., and Amaral, D. G. 2001 Increased social fear and decreased fear of objects in monkeys with neonatal amygdala lesions. *Neuroscience* 106, 653–658.

Preuschoff, K., Bossaerts, P., and Quartz, S. R. 2006 Neural differentiation of expected reward and risk in human subcortical structures. *Neuron* 51, 381–390.

Purves, D., Augustine. H. J., Fitzpatrick, D., Hall, W. C., LaMantia, A., McNamara, J. O., and Williams, S. M. 2004 *Neuroscience, Third Edition*. Sunderland, MA: Sinauer Associates.

Rachlin, H. 2000 *The science of self-control*. Cambridge: Harvard University Press.

Rachlin, H., Brown, J., and Cross, D. 2000 Discounting in judgments of delay and probability. *Journal of Behavioral Decision Making* 13, 145–159.

Raj, A. 2004 Overview and treatment of social anxiety disorder. *Manag Care* 13, 52–57.

Raleigh, M. J., McGuire, M. T., Brammer, G. L., Pollack, D. B., and Yuwiler, A. 1991 Serotonergic mechanisms promote dominance acquisition in adult male vervet monkeys. *Brain Res* 559, 181–190.

Rescorla, R. A., and Wagner, A. R. 1972 A theory of Pavlovian conditioning: variations in the effectiveness of reinforcement and nonreinforcement. In *Classical conditioning II: current research and theory* (ed. A. H. Black and W. F. Prokosy), pp. 64–99. New York: Appleton-Century-Crofts.

Rilling, J., Gutman, D., Zeh, T., Pagnoni, G., Berns, G., and Kilts, C. 2002 A neural basis for social cooperation. *Neuron* 35, 395–405.

Rogers, R. D., Tunbridge, E. M., Bhagwagar, Z., Drevets, W. C., Sahakian, B. J., and Carter, C. S. 2003 Tryptophan depletion alters the decision-making of healthy volunteers through altered processing of reward cues. *Neuropsychopharmacology* 28, 153–162.

Roiser, J. P., McLean, A., Ogilvie, A. D., Blackwell, A. D., Bamber, D. J., Goodyer, I., Jones, P. B., and Sahakian, B. J. 2005 The subjective and cognitive effects of acute phenylalanine and tyrosine depletion in patients recovered from depression. *Neuropsychopharmacology* 30, 775–785.

Roitman, M. F., Stuber, G. D., Phillips, P. E., Wightman, R. M., and Carelli, R. M. 2004 Dopamine operates as a subsecond modulator of food seeking. *J Neurosci* 24, 1265–1271.

Rolls, E. T., and Baylis, L. L. 1994 Gustatory, olfactory, and visual convergence within the primate orbitofrontal cortex. *J Neurosci* 14, 5437–5452.

Rolls, E. T., O'Doherty, J., Kringelbach, M. L., Francis, S., Bowtell, R., and McGlone, F. 2003 Representations of pleasant and painful touch in the human orbitofrontal and cingulate cortices. *Cereb Cortex* 13, 308–317.

Rustichini, A., Dickhaut, J., Ghirardato, P., Smith, K., and Pardo, J. V. 2005 A brain imaging study of the choice procedure. *Games Econ Behav* 52, 257–282.

Schneier, F. R., Blanco, C., Antia, S. X., and Liebowitz, M. R. 2002 The social anxiety spectrum. *Psychiatr Clin North Am* 25, 757–774.

Schoenbaum, G., Chiba, A. A., and Gallagher, M. 1998 Orbitofrontal cortex and basolateral amygdala encode expected outcomes during learning. *Nat Neurosci* 1, 155–159.

Schoenbaum, G., Chiba, A. A., and Gallagher, M. 1999 Neural encoding in orbitofrontal cortex and basolateral amygdala during olfactory discrimination learning. *J Neurosci* 19, 1876–1884.

Schultz, W. 2000 Multiple reward signals in the brain. *Nat Rev Neurosci* 1, 199–207.

Schultz, W. 2001 Reward signaling by dopamine neurons. *Neuroscientist* 7, 293–302.

Schultz, W. 2007 Behavioral dopamine signals. *Trends Neurosci* 30, 203–210.

Schultz, W., Apicella, P., Scarnati, E., and Ljungberg, T. 1992 Neuronal activity in monkey ventral striatum related to the expectation of reward. *J Neurosci* 12, 4595–4610.

Schultz, W., Dayan, P., and Montague, P. R. 1997 A neural substrate of prediction and reward. *Science* 275, 1593–1599.

Schultz, W., and Dickinson, A. 2000 Neuronal coding of prediction errors. *Ann Rev Neurosci* 23, 473–500.

Serrano, J. M., Iglesias, J., and Loeches, A. 1992 Visual discrimination and recognition of facial expressions of anger, fear, and surprise in 4- to 6-month-old infants. *Dev Psychobiol* 25, 411–425.

Smeltzer, M. D., Curtis, J. T., Aragona, B. J., and Wang, Z. 2006 Dopamine, oxytocin, and vasopressin receptor binding in the medial prefrontal cortex of monogamous and promiscuous voles. *Neurosci Lett* 394, 146–151.

Sugrue, L. P., Corrado, G. S., and Newsome, W. T. 2004 Matching behavior and the representation of value in the parietal cortex. *Science* 304, 1782–1787.

Tom, S. M., Fox, C. R., Trepel, C., and Poldrack, R. A. 2007 The neural basis of loss aversion in decision-making under risk. *Science* 315, 515–518.

Tremblay, L., and Schultz, W. 1999 Relative reward preference in primate orbitofrontal cortex. *Nature* 398, 704–708.

Tversky, A., and Kahneman, D. 1973 Availability—heuristic for judging frequency and probability. *Cogn Psychol* 5, 207–232.

Tversky, A., and Kahneman, D. 1981 The framing of decisions and the psychology of choice. *Science* 211, 453–458.

Vogt, B. A., Finch, D. M., and Olson, C. R. 1992 Functional heterogeneity in cingulate cortex: the anterior executive and posterior evaluative regions. *Cereb Cortex* 2, 435–443.

Westergaard, G. C., Suomi, S. J., Higley, J. D., and Mehlman, P. T. 1999 CSF 5-HIAA and aggression in female macaque monkeys: species and interindividual differences. *Psychopharmacology (Berl)* 146, 440–446.

White, N. M., and Milner, P. M. 1992 The psychobiology of reinforcers. *Ann Rev Psychol* 43, 443–471.

Winston, J. S., O'Doherty, J., Kilner, J. M., Perrett, D. I., and Dolan, R. J. 2007 Brain systems for assessing facial attractiveness. *Neuropsychologia* 45, 195–206.

Winston, J. S., Strange, B. A., O'Doherty, J., and Dolan, R. J. 2002 Automatic and intentional brain responses during evaluation of trustworthiness of faces. *Nat Neurosci* 5, 277–283.

Wise, R. A. 1987 The role of reward pathways in the development of drug dependence. *Pharmacol Ther* 35, 227–263.

Wise, R. A. 1998 Drug-activation of brain reward pathways. *Drug Alcohol Depend* 51, 13–22.

Wise, R. A. 2004 Dopamine, learning and motivation. *Nat Rev Neurosci* 5, 483–494.

Wood, R. M., Rilling, J. K., Sanfey, A. G., Bhagwagar, Z., and Rogers, R. D. 2006 Effects of tryptophan depletion on the performance of an iterated prisoner's dilemma game in healthy adults. *Neuropsychopharmacology* 31, 1075–1084.

Yarbus, A. L. 1967 *Eye movements and vision.* New York: Plenum.

Index

A

Accumbal dopamine, 53
Action potentials, tracking, 32
Alcohol craving, relapse prediction and, 137–161
  behavioral design of studies measuring
    neuronal systems activated by alcohol-
    associated cues, 144–148
      cue-induced brain activation and
        prospective relapse risk, 146–148
      important behavioral parameters in
        imaging studies, 145–146
  brain activity, core regions, 144–145
  brain-imaging studies, 137
  Clinical Institute Withdrawal Assessment for
    Alcohol Scale, 148
  conditioned alcohol craving, model of, 150
  conscious craving, 151
  cue-reactivity, 154
  dopamine D2 receptors, blockage of, 139
  dopamine dysfunction, 153
  dopamine self-firing, 152
  dysfunctional reward system, 154
  environmental confinement, 138
  GABAergic neurotransmission, 148, 149
  habitual drug intake, brain assessment, 147
  homeostatic counteradaptive process, 149
  impact of learning, 148–154
    clinical studies linking craving and
      relapse, 150–152
    computational models of phasic dopamine
      release, 152–154
  methodical approaches to study neuronal
    systems relevant for alcohol addiction,
    138–140
  neuronal network, model of, 141
  neurotransmitter dysfunction associated with
    altered cue reactivity, 144
  neurotransmitter systems implicated in
    craving for alcohol and other drugs of
    abuse, 140–144
  nicotine abuse, 151
  outlook, 155
  reward-associated error signaling, 142
  scope, 137–138
  sensory-motor gateway, 142
  serotonin transporter, 138
  using imaging to make clinical diagnoses,
    154–155
Alcohol cue effects, 7

B

Ambiguity, *see* Risk and ambiguity
AMP, *see* Amperometry
Amperometry (AMP), 105
Amphetamine, effects on dopamine signaling,
  112
Amygdala function, 196
Analog-to-digital conversion, 105

B

Background electrolyte (BGE), 118
BALB/c mice, 167
BDNF, *see* Brain-derived neurotrophic factor
Behavioral clamp, 37
Behavioral and molecular approaches in mouse,
  163–192
  BALB/c mice, 167
  C57BL/6J mice, 166, 169
  calcium-calmodulin kinase II, 171
  cell-specific genetic deletion in mouse self-
    administration studies, 183–187
  cell-specific transgenic expression in drug
    self-administration studies, 171–183
  direct striatal output pathway, 172
  dissociation of drug-taking and drug-seeking
    behaviors using self-administration
    models, 164–166
  dose-response curve, self-administration, 165
  doxycycline, 172, 173
  drug craving and relapse, model of, 166
  drug reinforcement schedule, 164
  dynorphin-containing striatal neurons, 182
  enkephalin-containing striatal neurons, 182
  extinction, drug seeking in, 166
  genetic manipulation, 168
  influence of genetic background on drug self-
    administration, 166–171
  in vivo target protein expression, 164
  low-dose sensitivity, 171
  molecular neuroadaptations, 163
  natural rewards, 174
  neurotransmitter receptors, 187
  potential for epistasis, 168
  self-administration model, 164
  serial testing procedure, 181–182
  tetracycline-inducible transgenic expression,
    172
  viral-mediated gene transfer, 184
BGE, *see* Background electrolyte
Biosensors

Printed and bound by CPI Group (UK) Ltd, Croydon, CR0 4YY

21/10/2024

01777044-0005